Information and Meaning in Evolutionary Processes

The most significant legacy of philosophical skepticism is the realization that our concepts, beliefs, and theories are social constructs. This belief has led to epistemological relativism, or the thesis that because there is no ultimate truth about the world, theory preferences are only a matter of opinion. Using evolutionary theory as the key to the naturalization of epistemology, William F. Harms seeks to develop the tools necessary to transform the philosophical study of knowledge into a proper scientific discipline.

This book will appeal to students and professionals in epistemology and the philosophy of science. William F. Harms is Research Associate at the Centre for Applied Ethics, University of British Columbia, and Philosophy Instructor at Seattle Central Community College.

CAMBRIDGE STUDIES IN PHILOSOPHY AND BIOLOGY

General Editor
Michael Ruse   *Florida State University*

Advisory Board
Michael Donoghue   *Yale University*
Jean Gayon   *University of Paris*
Jonathan Hodge   *University of Leeds*
Jane Maienschein   *Arizona State University*
Jesús Mosterín   *Instituto de Filosofía (Spanish Research Council)*
Elliott Sober   *University of Wisconsin*

Alfred I. Tauber   *The Immune Self: Theory or Metaphor?*
Elliott Sober   *From a Biological Point of View*
Robert Brandon   *Concepts and Methods in Evolutionary Biology*
Peter Godfrey-Smith   *Complexity and the Function of Mind in Nature*
William A. Rottschaefer   *The Biology and Psychology of Moral Agency*
Sahotra Sarkar   *Genetics and Reductionism*
Jean Gayon   *Darwinism's Struggle for Survival*
Jane Maienschein and Michael Ruse (eds.)   *Biology and the Foundation of Ethics*
Jack Wilson   *Biological Individuality*
Richard Creath and Jane Maienschein (eds.)   *Biology and Epistemology*
Alexander Rosenberg   *Darwinism in Philosophy, Social Science, and Policy*
Peter Beurton, Raphael Falk, and Hans-Jörg Rheinberger (eds.)   *The Concept of the Gene in Development and Evolution*
David Hull   *Science and Selection*
James G. Lennox   *Aristotle's Philosophy of Biology*
Marc Ereshefsky   *The Poverty of the Linnaean Hierarchy*
Kim Sterelny   *The Evolution of Agency and Other Essays*
William S. Cooper   *The Evolution of Reason*
Peter McLaughlin   *What Functions Explain*
Hecht Orzack and Elliott Sober (eds.)   *Adaptationism and Optimality*
Bryan G. Norton   *Searching for Sustainability*
Sandra D. Mitchell   *Biological Complexity and Integrative Pluralism*
Joseph LaPorte   *Natural Kinds and Conceptual Change*
Greg Cooper   *The Science of the Struggle for Existence*
Jason Scott Robert   *Embryology, Epigenesis, and Evolution*

Information and Meaning in Evolutionary Processes

WILLIAM F. HARMS
University of British Columbia

PUBLISHED BY THE PRESS SYNDICATE OF THE UNIVERSITY OF CAMBRIDGE
The Pitt Building, Trumpington Street, Cambridge, United Kingdom

CAMBRIDGE UNIVERSITY PRESS
The Edinburgh Building, Cambridge CB2 2RU, UK
40 West 20th Street, New York, NY 10011-4211, USA
477 Williamstown Road, Port Melbourne, VIC 3207, Australia
Ruiz de Alarcón 13, 28014 Madrid, Spain
Dock House, The Waterfront, Cape Town 8001, South Africa

http://www.cambridge.org

© William F. Harms 2004

This book is in copyright. Subject to statutory exception
and to the provisions of relevant collective licensing agreements,
no reproduction of any part may take place without
the written permission of Cambridge University Press.

First published 2004

Printed in the United States of America

Typeface Times Roman 10.25/13 pt.   *System* LATEX 2_ε [TB]

A catalog record for this book is available from the British Library.

Library of Congress Cataloging in Publication Data
Harms, William F.
    Information and meaning in evolutionary processes / William F. Harms.
      p. cm. – (Cambridge studies in philosophy and biology)
    Includes bibliographical references and index.
    ISBN 0-521-81514-2
    1. Knowledge, Theory of.   2. Evolution   I. Title.   II. Series.
BD177.H37   2004
121 – dc21   2003055312

ISBN 0 521 81514 2 hardback

Here then, is a kind of pre-established harmony between the course of nature and the succession of our ideas; ... As nature has taught us the use of our limbs, without giving us the knowledge of the muscles and nerves, by which they are actuated; so has she implanted in us an instinct, which carries forward the thought in a correspondent course to that which she has established among external objects; though we are ignorant of those powers and forces, on which this regular course and succession of objects totally depends.

David Hume, Enquiry, §V.

Contents

Acknowledgments	*page* xi
Introduction	1

PART I. GENERALIZING EVOLUTIONARY THEORY

1.	Replicator Theories	15
2.	Ontologies of Evolution and Cultural Transmission	52

PART II. MODELING INFORMATION FLOW IN EVOLUTIONARY PROCESSES

3.	Population Dynamics	81
4.	Information Theory	109
5.	Selection as an Information-Transfer Process	133
6.	Multilevel Information Transfer	150
7.	Information in Internal States	164

PART III. MEANING CONVENTIONS AND NORMATIVITY

8.	Primitive Content	189
9.	Is and Ought	218

Epilogue: Paley's Watch and Other Stories	241
Notes	247
Appendix: Proof of Information Gain under Frequency-Independent Discrete Replicator Dynamics for Population of n Types	253
References	259
Index	265

Acknowledgments

I would like to extend my grateful thanks to the many people who have played roles, small and large, in the development of the ideas herein, and in bringing this book to completion. These include Jeff Barrett, Michael Bradie, Werner Callebaut, Eric Cave, Peter Danielson, Peter Godfrey-Smith, Karl Hufbauer, David Hull, Libby Logerwell, Ned McClennen, Ruth Millikan, David Shoemaker, Elliot Sober, Kim Sterelny, and Bill Wimsatt. Special thanks go out to Michael Ruse for suggesting this book in the first place, and seeing it through to the end; and to Brian Skyrms for showing me the power of dynamical models in philosophy.

Introduction

WHY EPISTEMOLOGY MATTERS

The nagging desire not only to gain knowledge but to understand the nature of knowledge itself arises in different ages for different reasons. The justification of religious truths and ethical precepts, the desire for certainty in an uncertain world – both have been effective at motivating inquiry into the fundamental nature of knowledge. In our era, a story can, perhaps, best illustrate the need. The story I have in mind involves a large rat trap. It also involves two young university professors, a celebrated theory war, about eight hundred undergraduate students, and a book by Thomas Kuhn. It is a true story, or at least it started out as a true story. This is how I remember it.

Young Professor B and young Professor S had been recruited from their respective departments, philosophy and German literature, to participate in an interdisciplinary extravaganza known as "Core Course." Core Course was the sort of politically correct Great Books program common to large universities. Drudge work for the eight hundred students (assignments, grading, one-on-one contact, and daily class time) was handled by twenty or thirty "teaching associates" (second-year graduate students), staff meetings once a week. Young Professors S and B were accorded starring roles in this production, giving primary content lectures to the eight hundred students in two shifts, the auditorium being limited to half that number. In their home departments, Professor B's job was to do something called "philosophy of physics," which involved a lot of mathematics, and to teach undergraduates things such as logic and critical thinking. Professor S's job was to interpret great works of German literature, to teach students about this, and to teach things such as postmodernism and deconstruction. Needless to say, Professors B and S disagreed about some very basic things. This was the winter quarter session, which included material on the history and philosophy (and sociology) of science.

The incident of interest occurred during the lectures on Kuhn's influential *The Structure of Scientific Revolutions*. In case you've forgotten, this is the book that got everyone talking about "paradigm shifts" in the 1960s. Kuhn's basic thesis was that scientific progress cannot ever be the steady accumulation of truths over centuries that its proponents sometimes claim. Instead, science progresses through a series of lurching cycles driven by the acquisition of various bits of scientific infrastructure – concepts, measurement apparatuses, and model studies – followed by the discovery of the limits of that technology. This results in a cycle of crisis and revolution in which the old infrastructure is replaced by new, frequently accompanied by an actual loss in explanatory power for the discipline in question. Critics of science had a field day with this stuff, much to Kuhn's dismay, claiming that he had shown that scientific objectivity is a mere social and institutional fiction, science having no more inherent value than witchcraft or performance art. This is, of course, not what Kuhn had in mind, but that's a different story.

Professor S had first go at this material and took the opportunity to promote some of his favorite causes – that reality is a social construction, that political power drives the content of all belief systems, and that while the science folks seem to have been getting a lot of the attention (not to mention money) lately, what this means is simply that they have the power, no more. Lore is lore. The Upanishads and Newton's *Principia* are on a par in this respect.

Young Professor B could not take this lying down. So when his turn to lecture came, he showed up with the aforementioned rat trap, his intent being to challenge Professor B to a sort of practical duel of ideologies. The lethal efficiency of the rat trap was demonstrated – lethal to a pencil, anyway – and Professor S was offered the opportunity to demonstrate the much-vaunted social construction of reality. Surely, if Professor S truly believed reality to be a construction, then he should be quite happy to put his finger in the trap, secure in the knowledge that the trap could not *really* hurt him if he believed it would not. Quick on his feet, Professor S replied that he was sufficiently at the mercy of the way his own society constructs reality that he was quite sure that the trap would actually hurt him. But, he insisted, someone from a society which constructed reality in a sufficiently different way might not be harmed. With this impasse, it was time for the bell.

To anyone who is even marginally aware of what has been going on in the humanities at major U.S. universities during the last forty years, the theoretical conflict that lies behind this incident is familiar and, if anything, the incident itself was relatively free from the acrimony one often finds in less-public conversations on the matter. Analytic philosophers can be viciously critical of the "fuzzy-headed thinking" in so-called continental

Introduction

philosophy and of the destructive potential of unchecked relativism, historicism, and deconstruction. Continental philosophy, usually a bigger force in the humanities at large than in philosophy departments proper, views analytic philosophers as insufficiently aware of the social, historical, and political contexts in which humans live and of the effect these contexts have on how people understand themselves and their world. The left claims that the right is in denial about the constructedness of reality. The right claims that the left has given up its right to a place in an institution dedicated to the furtherance of human understanding.

As is usual in these kinds of disagreements, both sides have a point. On one hand, human knowledge is not discovered but constructed. The beliefs we have are constructed from the concepts available. Those concepts are in turn constructs, and while one might be able to argue that, for a given purpose, some set of concepts is optimal, there seems little reason to think that one set of concepts is optimal for all purposes of all species at all times. On the other hand, even accepting the constructedness of our conception of reality, one may reasonably insist that not all such constructions are on a par from a practical point of view and that human beings and the things they do are sufficiently similar that some worldviews might well be flat out better than others.

Ideally, one would think that this crucial philosophical question about the status of knowledge would be the proper subject matter of epistemologists and that epistemology as a discipline would step up to clarify these matters. This has not been the case. Outside of analytic philosophy many believe, as Rorty (1979) argued, that the death of foundationalism has left epistemology as an impossible and unnecessary discipline, given that epistemologists have traditionally attempted to discover some area of human belief that transcends the possibility of doubt. Meanwhile, epistemology itself seems to have become something that philosophers do "on the side" when epistemological issues arise in their areas of primary interest or something done by specialists who work on small areas of the large epistemological puzzle. Looming is the question of how, in the face of the constructedness of our knowledge of reality, to say anything non–question begging about the relationship between that conception and the world we want to believe it mirrors.

WHY EVOLUTION MATTERS

What, if our conception of reality is a construct, can epistemology possibly salvage? Two possibilities come to the fore. First, we ordinarily have a certain amount of confidence that the care we take in acquiring and evaluating beliefs is not a complete waste of time. It generally seems to us that beliefs have something to do with the reality "out there," and ordinary perception and common sense are largely to be trusted. What reason can we have to trust our senses, minds, and memories? What reason can we have to think that our cognitive and perceptual apparatus is reliable? Such questions suggest themselves for our consideration even if we accept that certainty is not to be had and even if we acknowledge that there is probably no such thing as *the* truth for all people at all times.

Second, "knowledge" itself is an evaluative, normatively loaded term, and our conception of knowledge is replete with a number of such concepts. Truth, justification, and meaning all demand explanations. The very concept of knowledge implies more than just that some mental states reliably and usefully track the world. It implies that there is such a thing as getting it right, that there are some sort of rules which apply to how we form and interpret beliefs that go beyond mere usefulness. Moreover, it is not only the rules regarding knowledge that we claim to know. We claim to know the difference between right and wrong, just and unjust. We claim that there is something to *know* about such matters, but what might ground such knowledge is more of a mystery than the basic question of how our concepts and beliefs relate to the world. At least in the latter case, we can ask how concepts relate to whatever is "out there" that makes things happen. In the case of the normative question, the various sorts of "oughts" or rules do not even seem to be a matter of what is going on with the "hidden springs and principles" of nature, to borrow Hume's phrase. The purported rules are a matter not of what is, but of how things ought to be.

Together, these two questions concerning reliability and normativity present the basic challenge for epistemology. In this book I try to show that, in both cases, the answers come from understanding evolutionary processes.

In the first case, the general solution is not so hard to see, nor is the suggestion particularly novel. Briefly, the only reason to think that our thoughts usefully and reliably mirror reality is that if they didn't at least get us coping fairly well with the world, it's not likely we would be here at all. The principle we are alluding to is *natural selection,* the driving force of biological evolution, and the school of thought that emphasizes the importance of natural selection for the understanding of knowledge is usually called *evolutionary epistemology* (Campbell 1974).

Evolutionary epistemology differs from other sorts of naturalistic epistemology in the extent to which it emphasizes the importance of natural selection. "Naturalistic" epistemologies in general eschew the traditional emphasis on introspection (as well as the traditional acceptance of at least the

Introduction

possibility that minds are more than mere matter) for the scientific point of view. For naturalists, human beings are material entities and anything that is of interest about them, including the phenomenon of knowledge itself, is to be studied empirically and scientifically. This in itself does not necessarily bring with it any commitment to the importance of evolution. On the contrary, classic naturalistic epistemologists such as Quine (1969) insisted that epistemology was to consist of observing actual physical processes involved in perception, cognition, and the like, and while he noted that evolutionary history was responsible for the similarity of our quality space to the quality space of the world, reasoning about evolutionary history in any detail did not seem to be part of his program for the new epistemology. Evolutionary epistemologists, while also naturalists in their insistence on the status of human beings as material entities, believe that we cannot fully understand knowledge purely in terms of the currently observable causal processes on which the physical sciences focus. This is not to say that one needs any more than an understanding of current causal structures and processes to understand *how* we learn about our world. But understanding *why we are able* to so learn requires some understanding of our evolutionary history. Moreover, given that epistemology must ultimately attempt to give a coherent account of its own nature, understanding the general principles of how evolution adapts cognitive systems to environments seems essential to understanding the *nature* of the relationship between thoughts and the world. Even for those of us who have given up on foundationalism, some way of standing back and thinking about the fit between our own concepts and the world, as well as a way of understanding comparisons between different worldviews, is desirable. If we can understand something about how much and what kind of optimization of cognitive fit is likely to emerge from evolutionary histories, we will be that much closer to a more adequate scientific understanding of knowledge and more able to answer some of the interesting comparative questions about knowledge that arise between cultures and between species. The general idea of an evolutionary epistemology has been around for some time. The pressing need at the current juncture is for the development of theory and, in particular, formal tools of analysis that will make the relationship between selection and mind-world coordination more tractable.

As for the second question, evolutionary epistemologists have, for the most part, adopted the naturalistic party line in abandoning the traditional normative concerns of epistemology, concerns with capitalized topics such as Truth, Justification, and Reason. Hume, possibly the first modern naturalistic philosopher, argued that judgments of *moral* propriety were simply the operation of certain sentiments. Early-twentieth-century naturalism resulted in the similar thesis of emotivism, according to which normative value judgments are merely the expression of emotions, devoid of the kind of referential content that might make them objectively true or false. In this case, what applies to morality applies to Truth, Justification, and Reason as well. Indeed, the current majority opinion seems to be that naturalism unavoidably brings with it the reanalysis of any and all normative judgments in terms of the proximal emotions that give rise to them. The sting of abandoning so much of the traditional concern of epistemology may be salved to some degree by the thought that, after all, it does no good to complain that the things people in fact do are not the things they *ought* to do. On the contrary, the more energy we spend on understanding both how and why people do what they do, the sooner we will be able to stop complaining and actually do something about human epistemic failings. Still, the most common reason for rejecting naturalistic epistemology is its alleged failure to accommodate the normative, and proponents of naturalistic epistemologies have had little to say in opposition (Kim, 1988). It would seem that either naturalism is incomplete or the normative is to be ignored along with the rest of the supernatural. In either case, one won't be able to get an *ought* from an *is*. The current philosophical consensus seems to be that if both is (facts) and ought (norms) are to be studied, they will be studied in fundamentally different ways, perhaps by very different disciplines. This consensus also maintains that all previous attempts to explain value on the basis of fact have failed. As a result, anyone attempting to account for the truth of normative judgments within a naturalistic framework has to contend not only with the inherent difficulty of the task, but also with the fact that everyone already *knows* it can't be done.

I argue that the problem with our current understanding of the gap between *is* and *ought* is that it is based on an outdated conception of the proper domain of scientific enquiry. Taking physics as the paradigm, we have conceived of science as concerned exclusively with occurrent causal processes. Like many such conceptions, this one has been subject to considerable erosion in the centuries since its emergence. Just as seventeenth-century mechanistic materialism was forced to broaden itself to accommodate "action at a distance," first in the form of gravity and then in the form of electromagnetism, so the empiricist insistence that contingent history is merely anecdotal is gradually giving way to the recognition of the central importance to biology of essentially historical relationships like kinship and adaptation. The increasing importance of historical relationships is in turn driven by the increasing status of biology as a genuine science and the concomitant change in our conception of what constitutes a genuine science. It turns out that these historical relationships pretty much make a shamble of the arguments of Hume and Moore,

Introduction

which seem to lie behind the common "knowledge" of the intractability of the is-ought gap.

Exactly how the historical subject matter of evolutionary biology changes the big picture will be the subject of the final section of this book. Roughly, I argue that a broad enough theory of meaning can account for the meaning and thus truth of both "is-statements" and "ought-statements," and in so doing, account for the difference between them. In the absence of traditional assumptions regarding abstract meaning-entities (e.g., propositions), we are forced to recognize that meaning is conventional, and conventions are historical entities. This focus on the ways in which meaning conventions emerge highlights facts about meaning that are usually overlooked. In the end, the general analysis of the emergence of meaning conventions shows indicative language to be a rather specialized sort of signaling system, and not the only one capable of correspondence truth. Normative utterances are more akin to warning cries than to statements of fact, untranslatable but not unanalyzable in our usual mode of descriptive speech.

Given the general consensus that this sort of project cannot work, it behooves me from the start to be completely clear about my own aspirations with respect to a theory of the normative. I argue that a full knowledge of the functional history of signaling systems of all sorts, including especially the system-stabilizing consequences of signaling behavior, is sufficient to establish the conventions governing meaning and truth for all sorts of signals, from hormonal secretions to scientific hypotheses to pronouncements of moral and epistemic justification. The point, however, is purely academic in that the sorts of historical facts relevant to determining meaning-conventions are so difficult to come by that we should not ever expect to see our own intuitions on such matters overridden by pronouncements of evolutionary science. There is, in addition, a purely theoretical reason why you cannot get an ought from an is, but we will be in position to actually explain this, rather than merely recognize it as we usually do. So, even if the possession of all the relevant facts allowed us to establish the meaning of normative utterances and thus tell us whether they are true, this is not quite the same as *telling* us what we ought to do. As a consequence, both according to the strict letter of the theory and because of the unavailability of pertinent facts, the theory I propose falls short of the traditional aspirations of epistemologists – to issue authoritative epistemic norms. I am enough of a naturalist that this doesn't bother me. What I do think is philosophically important is the way in which a defensible descriptive theory of the normative can counteract the relativism with respect to normative standards which has accompanied our materialistic worldview. If a theory can reassure us that, even in a purely material world, there may be standards of conduct and thought that are sufficiently general and objective to apply to *all* human beings, this is no small thing. Indeed, by my lights, we are far more in need of a theory which helps us make sense of the very possibility of objective norms than we are of one which tells us what they are.

ONTOLOGY, SELECTION, AND CONVENTION

The nine chapters of this book are organized into three parts. Parts I and II are concerned with furthering the general evolutionary epistemology project: Part I with ground clearing, Part II with the construction of formal tools for analyzing multilevel selection and information-transfer processes. Part III deals with the more contentious evolutionary take on meaning conventions for normative intuitions. The three parts can be read independently and may appeal to different interests and temperaments.

Part I

Evolutionary epistemology, at least in its more ambitious versions, requires a way of thinking about evolution that is broad enough to allow it to occur in culture as well as in the genetic lineage. The most common such conception is Dawkins's (1976) "meme," a self-replicating "informational entity." Dawkins, along with a number of others, is convinced that evolution by natural selection only happens to lineages of self-replicating entities; consequently, if evolutionary concepts are to apply to culture, then there must be some sort of self-replicating entity in culture.

In Chapter 1, I discuss at some length versions of the cultural replicator from Dawkins, Hull, and Dennett. I conclude that there are not one but three notions and that none is adequate for epistemological purposes. In Chapter 2, I broaden the critique to the nascent field of memetics in general, fielding alternatives and trying to set some standards for conceptual innovation. I argue that genes are not the only or even the best way of thinking about the tree of life – the lineage of dividing cells has a concrete identity over time that genetic "information" does not. Cells, moreover, may be the only thing that can truly be said to "self"-replicate, and the logic of that process is fission and regrowth rather than transcription. Consequently, looking for genelike entities as a prerequisite for cultural evolution is rather misguided. Nor does cultural transmission require an entity to be transmitted. On the contrary, communication and cultural transmission are easily understood as coordinated state change in closely related organisms. Finally, even if memetics is defended

Introduction

only as a way of looking at cultural evolution and transmission, there are limitations to its utility.

Part II

Chapter 3 begins my positive account, developing a general model of evolution derived from the formal models of population genetics and evolutionary game theory. The emphasis here is twofold: first, on extracting general evolutionary concepts such as selection, fitness, and variation from our best abstract mathematical understanding of evolution rather than proposing causal analogs to the actual physical process of biological evolution as the replicator approach does; and second, on generating a simplified formal model suited for computer modeling of both biological and cultural evolution. Chapter 4 explores the mathematical concept of mutual information for the purposes of epistemology and establishes some simple results regarding the utility of information. Chapter 5 puts the two together, establishing natural selection as an information transfer process. Chapter 6 develops a two-level selection and variation model interpreted as a model of bacterial navigation, with an eye to creating a formal model for the basic interdependencies between biological and cultural evolution. Chapter 7 develops a three-level model of bumblebee foraging, which accommodates the formation of preferences and measures information about the environment in these characteristically *internal* states. Because the information transfer-model in Chapter 7 is technically rather difficult, I try to explain its implications in simpler terms at the end.

Part III

To reiterate, even a purely descriptive account of normative intuitions and language can help constrain cultural relativism with respect to them. Chapter 8 develops a model of primitive meaning content, inspired by the teleosemantic theory of Millikan (1984) and a game theoretic model from Skyrms (1996). This general model is then applied to regulatory hierarchies, resulting in systems with multiple semantic maps which share many formerly puzzling features of human normative deliberation. Chapter 9 defends the plausibility of this primitive-content hypothesis against standard objections. In particular, I show that the "open question argument" of Moore and Hume's famous analysis of normative relationships do not apply to historical-functional semantic theories along with a baker's dozen (or so) other philosophical worries. The reason you can't get an ought from an is is explained.

Throughout, I try to keep focused on the theoretical objectives involved in an evolutionary theory of knowledge. This means resisting the temptation to delve into "hot" topics such as the nature of consciousness, group selection, the excesses of adaptationism, the marginalization of developmental biology, and the naturalization of ethics. I apologize in advance to readers who find this stinginess unsatisfying.

PREESTABLISHED HARMONY

German philosopher Gottfried Wilhelm Leibniz (whose most famous accomplishment may well be the invention of the differential calculus simultaneously with Newton in the closing years of the seventeenth century) argued that substances exist but do not interact, each separate thing sufficient in itself to determine its unfolding over time in the absence of causal interactions with other things. One consequence of this view was that our knowledge of things in the world cannot be the result of causal interactions with them, as we usually assume. Leibniz's solution to this epistemological difficulty was that God, in creating each thing ("monad"), had done so in such a way that the unfolding of each thing over time was mirrored by others. Thus, perception and knowledge are not the result of causal interaction but of a "preestablished harmony" between the movements of our minds and the movements of things in the world.

Scottish philosopher Hume, like Leibniz, called into question our ordinary notion of causality, but as an empiricist, both his method and his conclusions were different. Hume asked us to consider closely not conceptual analysis but the act of perceiving causal interactions. He simply pointed out that all we ever see are sequences of events, never the actual force of one object acting on another. Causal power is something we project onto interactions, rather than perceive in them, and the notion of causality itself, he speculated, was merely the generalization of our own expectant impulses. Despite or, perhaps, because of Hume's informality and reliance on common sense, his argument remains to this day *the* problem of causality, a large part of his not inconsiderable philosophical legacy. Along with the "problem of induction," his argument concerning causality cemented his reputation as one of the greatest skeptics and philosophical troublemakers of all time. What is usually disregarded is his own proffered solution to the difficulties he raised.

Hume argued that causal and inductive reasoning, along with moral judgment, are not the result of the rational perception of eternal laws of reason and standards of behavior but are in each case the result of instinct, habit, or sentiment. In contemporary terms, Hume believed that we are just "wired" to think the way we do and that rationalist notions of being able to perceive certain relationships as self-evident by the "light of nature" are just nonsense. This was not intended to undermine our reliance on causal or inductive reasoning nor our genuine belief in moral standards, however, but to pull the rug out from under rationalist pretensions that Reason was the final authority on all aspects of human thought and behavior. But how, without recourse to God or the power of Reason, were we to establish the propriety of causal and inductive inference?

Hume did know when to give it a rest. Philosophy, he said, takes one into distressingly deep waters and forces one to conclusions that seem to undermine everything one believes. Fortunately, human nature is too strong for mere philosophical arguments to freeze us into inaction, and one needs to know when it is time to put down the philosophy and go play billiards. This pragmatic streak shows itself in this "solution" to his skeptical doubts as well. It may be that causal and inductive inference are merely the operation of instincts or habits. Nonetheless, any fool can see that they are *good* instincts and habits, without which we would be incapable of surviving. Causal and inductive reasoning seem to "fit" with the patterns of the world in just the right way to help us cope with them. How this could be so was not clear, but that it is so only a philosopher could doubt. What he said was this:

> Here, then, is a kind of pre-established harmony between the course of nature and the succession of our ideas.... As nature has taught us the use of our limbs, without giving us the knowledge of the muscles and nerves, by which they are actuated; so has she implanted in us an instinct, which carries forward the thought in a correspondent course to that which she has established among external objects; though we are ignorant of those powers and forces, on which this regular course and succession of objects totally depends. (Hume 1739/1978, §V)

The reason for bringing up Hume's little joke on Leibniz is that Hume was right. Nature *has* implanted in us cognitive instincts which keep our thoughts in productive harmony with the world, and this is the secret to understanding knowledge. What we know and Hume could not have is how, in fact, nature has gone about establishing this harmony between cognition and worldly processes. One hundred years before Darwin, Hume's confidence in our reasoning processes was merely common sense, without theoretical foundation. For us, more than two centuries later, a great deal is known about both the broad outlines of evolutionary theory and the details of our own evolutionary history. Much work remains to be done in both areas, and the empirical task of working out the details of our evolutionary history may well never be completed, due to both the enormity of the task and the antiquity of the relevant facts. Theory can, at least temporarily, provide a little more closure.

This book, then, is a contribution to the theory behind Hume's preestablished harmony. Insofar as it is successful, it does what successful philosophy often does – takes its subject matter a little closer to being an autonomous science with its own distinctive methods, a little farther from being the proper domain of philosophers.

I

Generalizing Evolutionary Theory

1

Replicator Theories

Their Proponents and Limitations

The idea that culture evolves in a Darwinian way via variation and selection has been around for some time. Evolutionary epistemology, at least in its more ambitious versions, critically depends on the theoretical defensibility of the notion of cultural evolution, although to date there is considerable difference of opinion as to what exactly it is that evolves over cultural history. Current theoretical frameworks for understanding culture this way fall into two categories. The first approach reasons analogically from biological evolution. The gene is taken as the essential ingredient to biological evolution, and it is reasoned that if there is anything sufficiently like a gene in culture, then Darwinian explanations (and expectations) can apply to culture as well as basic biology. In short, Darwinism can apply to culture just in case there is a cultural "replicator."

The purpose of this chapter is to examine the notion of a replicator, the attempts to apply it to culture, and its shortcomings as a central notion for evolutionary epistemology. This discussion accomplishes several things. It introduces the reader to the most notable attempts to construct an evolutionary theory of culture and to their general level of sophistication. I also spend a considerable amount of time examining their failings. This will demonstrate the pressing need for a nonreplicator model of evolutionary processes, the construction of which is the main task of Part II.

As we will see in Chapter 3, many evolutionary biologists understand that one need not have any sort of replicator to have Darwinian evolution, and typically formal models of evolutionary processes require no such entity; and this includes the so-called replicator dynamics.[1] As such, it is of some interest to account for the prevailing assumption that evolutionary processes do need replicators. We begin by taking a look at Richard Dawkins's writings, particularly his (1976) *The Selfish Gene*, which brought to the thinking public a compelling vision of evolutionary processes centered around the notion of a replicator and which seems to be generally responsible for the popularity of the notion. More important for our purposes here, Dawkins was also the proximal author of the notion of a "meme," a sociocultural replicator, and this was the prototype of the replicator which now so commonly finds its place in evolutionary accounts of culture.

David Hull's (1988a) *Science as a Process* stood for some time as the benchmark for evolutionary accounts of science, and Hull's approach to understanding evolution has had considerable influence on the new generation of philosophers of biology. Consequently, it is worthwhile to examine in some depth Hull's theory and its shortcomings for the purposes of epistemology. Finally, the current popularity of the meme is probably due to the efforts of Daniel Dennett, who proposes a rather different semantic or information classification of these elusive entities. I argue that there is not one but three notions of the sociocultural replicator, none of which is adequate to the purpose of a rigorous science of cultural evolution.

DAWKINS, REPLICATORS, AND MEMES

The Selfish Gene (1976) had both a purpose and a vision. The purpose was the promotion of Dawkins's "gene selectionist" version of biological evolution, directed largely toward putting an end to a lot of bad appeals to evolutionary explanation, particularly invocations of group-selection arguments (i.e., what evolves is what is good for the group) from people outside of the biological sciences. To accomplish this, it had to attempt to be both accessible and scientifically respectable, and this it is. Although almost entirely devoid of technical language, it is basically good biology and clearly presented. For it to have had the vast influence that it has, especially with the lay public and with those outside of the biological sciences who are interested in biological evolution, it had to be compelling as well. It had to offer an image that would stick in the mind. This is where the "vision" comes in. It goes something like this.

> The replicators that survived were the ones that built *survival machines* for themselves to live in. The first survival machines probably consisted of nothing more than a protective coat. But making a living got steadily harder as new rivals arose with better and more effective survival machines.... Four thousand million years on, what was to be the fate of the ancient replicators? They did not die out, for they are past masters of the survival arts. But do not look for them floating loose in the sea; they gave up that cavalier freedom long ago. Now they swarm in huge colonies, safe inside gigantic lumbering robots, sealed

off from the outside world, communicating with it by tortuous indirect routes, manipulating it by remote control. They are in you and in me; they created us, body and mind; and their preservation is the ultimate rationale for our existence. They have come a long way, those replicators. Now they go by the name of genes, and we are their survival machines. (Dawkins 1976, 19f)

What this image does, along with the many like it scattered throughout the text, is to make Darwinian evolution thinkable for the vast majority of us nonspecialists. These blindly replicating strands of nucleic acid and their "lumbering robots" stick in the mind in a way that the formulas of population genetics and notions such as *meiosis* and *mitosis* do not. We need not feel guilty about making use of these sensationalistic science-fiction images, because the biological theory behind them is generally sound. The problem, however, is that replicators are a central part of this image, and if this is how we have learned to think about Darwinian processes,[2] then it is natural for us to assume that such processes must have replicators at their heart.

Darwin's (1859) theory of evolution by natural selection had no replicators, at least not in the current sense (although it did have "gemmules" of reproductive information). The emphasis was on populations of organisms, related by descent, whose Malthusian tendency toward geometric increase in population size led inevitably to competition for limited resources, and thus to the "struggle for existence." The focus was on what we now call the phenotypic individual, and on the species. Neo-Darwinism, as it has emerged following the "modern synthesis," finds a different focus. Following the merging of evolutionary theory and genetics by Dobzhansky in his (1941) *Genetics and the Origin of Species*, the discovery of the physical means of genetic transmission – DNA, and the gradual realization that most evolutionary change is a matter of relatively subtle and gradual changes in the distributions of traits in populations in response to complex and probabilistic "selection pressures," population genetics has found its place at the core of evolutionary theory. These days we theorize evolution in terms of mathematical tools which model the evolving frequencies of genetic fragments within populations, and it is these on which natural selection is commonly thought to operate, at least on the smallest scale. It is these fragments, whichever ones selection actually operates on, that we call genes, and these are also the replicators of Dawkins's vision.

The strength of Dawkins's discussion of replicators lies in his treatment of the more specific case of genes and of the question of which kinds of biological entities can function as units of selection, entities which natural selection is selecting for or against. But since we are going to be worrying about whether entities other than biological ones are replicators, however, and whether one always needs replicators for Darwinian evolution, we need a definition of the more general notion of "replicator." Unfortunately, Dawkins's discussion tends to drift back and forth between the general case of replicators and the more specific biological cases (indeed, he tends to use the terminologies interchangeably), and much of the discussion of the criteria for replicators occurs in the context of "units of selection" arguments. So, a bit of detective work is in order.

The place to begin is with the notion of stability. Dawkins writes,

> Darwin's "survival of the fittest" is really a special case of a more general law of *survival of the stable*. The universe is populated by stable things. A stable thing is a collection of atoms that is permanent enough or common enough to deserve a name . . . or it may be a class of entities, such as rain drops, that come into existence at a sufficiently high rate to deserve a collective name, even if any one of them is short-lived. (Dawkins 1976, 12)

There are two things to note here. One is the notion of stability. Evolution is a matter of the persistence and proliferation of stable things. Adaptation to a given environment is a matter of the increased stability of (e.g.) an organism in that environment. We will see that stability conditions are central to Dawkins's criteria for replicators and for units of selection. But the stability of what? This is the second point. We should not be distracted by the vagueness of "collective entities" which "deserve a name." The point is that the replicators Dawkins wants to talk about, genes, are types (of a rather particular sort) rather than tokens. As such, the stability conditions involved need to be applicable to types of entities as well as their tokens. Gene tokens of a given type form a collective entity in this sense, defined by their identical chemical structure, and the manner of their production, replication, ensures that they can "come into existence at a sufficiently high rate" for the populations of tokens to be stable. Notice that different types of criteria are allowed for "collective entities" or types. Unlike genes, the type "raindrops" is not defined by chemical structure, but by something like composition (H_2O), cause (atmospheric precipitation), size, and perhaps downward velocity. Later, we will be concerned with functionally defined types, such as the type "brain-states that cause their holders to act as though they believe X." In each case, we are concerned with the conditions that account for the stability of the types and for the proliferation of the tokens.

It is against this backdrop that replicators are introduced.

> At some point a particularly remarkable molecule was formed by accident. We will call it the *Replicator*. It may not necessarily have been the biggest or the most complex molecule around, but it had the extraordinary property of being

able to make copies of itself.... Actually, a molecule that makes copies of itself is not as difficult to imagine as it seems at first, and it only had to arise once. Think of the replicator as a mold or a template. (Dawkins 1976, 15)

The replicator we have here is a type of molecule. It is the ancestor of our own DNA, and its distinguishing mark is that it achieves its great stability through time by making copies of itself. Even though its individual tokens may have a relatively short life span, the collective entity determined by their common chemical structure is "immortal," or at least potentially so.

The passage just cited is about as close as we ever come to a definition of "replicator" in *The Selfish Gene*. Fortunately, Dawkins is a bit more forthcoming in his *The Extended Phenotype* (1982). There he writes,

> I define a *replicator* as anything in the universe of which copies are made. Examples are a DNA molecule, and a sheet of paper that is xeroxed. Replicators may be classified in two ways. They may be "active" or "passive," and, cutting across this classification, they may be "germ-line" or "dead-end" replicators.
>
> An *active replicator* is any replicator whose nature has some influence over its probability of being copied.... A *passive replicator* is a replicator whose nature has no influence over its probability of being copied...
>
> A *germ-line* replicator is a replicator that is potentially the ancestor of an indefinitely long line of descendant replicators.... A *dead-end replicator* is a replicator which may be copied a finite number of times, giving rise to a short chain of descendants, but which is definitely not the potential ancestor of an indefinitely long line of descendants. Most of the DNA in our bodies are dead-end replicators. They may be the ancestors of a few dozen generations of mitotic replication, but they will definitely not be long-term ancestors. (1982, 83)

The evolution of life, then, is just the proliferation of active germ-line replicators, and this is Dawkins's reason for introducing these distinctions. Even the definition offered here is not adequate for our purposes, however, since we are still not clear on what makes a replicator. On one hand, it seems that anything at all could be a "passive" replicator. All that is necessary for this is that it be subjected to some sort of copying process. A passive replicator is not a kind of thing, but a role that something plays in a larger process. Active replicators, on the other hand, are clearly intended to be things that can be said to "make copies of themselves" in the appropriate environment. It is important to note that Dawkins is being much more cautious here than he was in the passage from *The Selfish Gene* in specifying the importance of environmental conditions for replication. We worry about this question later, whether the difference between replicators that "make copies of themselves" and those that "have copies made of them" is relevant to how we go about modeling Darwinian processes. The reason that most of our DNA consists of "dead-end" replicators is that they occur in somatic cells, whose lineages must die when the organism dies.

The more interesting theoretical factor limiting what can function interestingly as a replicator is that "crossing-over" or recombination in gamete production tends to break up the longer segments of nucleic acid with high enough frequency to offset the effects of selection on those segments, and it is just these longer segments that Dawkins is talking about. This is a matter of the "longevity" of a replicator, and this notion is central to Dawkins's discussion. Dawkins (1976) is primarily concerned with the conditions under which something can function as a replicator in an evolutionary process. Longevity in generation time is one of these conditions or criteria. He writes,

> Individuals are not stable things, they are fleeting. Chromosomes too are shuffled into oblivion, like hands of cards soon after they are dealt. But the cards themselves survive the shuffling. The cards are the genes.... They are the replicators and we are their survival machines. When we have served our purpose we are cast aside. But genes are denizens of geological time: genes are forever. (1976, 35)

This failing of individuals, that is of biological organisms, as replicators is common to populations also and, as I noted earlier, this latter concern with group-selection thinking seems to be Dawkins's primary motivation in the search for proper units of selection. If one admits that phenotypic individuals are replicators and can be operated on by selection, then it is just a short step to group-selectionist thinking. But we should not be misled here. It is not that individuals are tokens rather than types that keeps them from functioning as replicators, but rather that the types of which they are tokens are not unitary enough to function as units of selection.

The issues of unity or identity of tokens of the type and the longevity of that type as an entity are closely intertwined, but the importance of both regarding the functioning of Darwinian processes is this. Evolution by natural selection is generally not a matter of survival versus obliteration, but rather is a matter of small changes in the frequency of replication due to subtle interactions with the environment which affect that frequency. For an entity to be selected by the environment in preference to something else, its interaction with the environment must be consistent over many generations. This means that whether something is a replicator, even an "active germ-line" replicator, is relative to its generation time, as well as to other factors.

These considerations dictate Dawkins's handling of the definition of "gene," at the same time making the task of determining just what a replicator is more difficult.

The one thing to be clear about before proceeding to the specifics of the definition of gene is that "On any definition, a gene has to be a portion of a chromosome" (Dawkins 1976, 28f). What we are talking about here is a segment of a strand of DNA. What makes two strands contain the same gene is that they contain segments which have an identical chemical structure.[3] The genetic code is discrete, and this discreteness contributes to genes being the kind of "immortal entities" that they are in two ways: (1) copying error is reduced, increasing the stability of the types, and (2) we are justified in talking about genes as well-defined types, even when we don't know their exact chemical structure, since they are differentiated according to a discrete code. There is no theoretical problem with the identity of two segments of chromosome (although when two genes are the same is still a bit of a puzzle), and this fact is central to the role they play in biological evolution, in both (1) the process itself, and (2) the development and application of the theory. Just how much of a chromosome is a gene, then?

> I shall make no attempt to specify *exactly* how long a portion of chromosome can be permitted to be before it ceases to be usefully regarded as a replicator. There is no hard and fast rule, and we don't need one. It depends on the strength of the selection pressure of interest. We are not seeking an absolutely rigid definition, but "a kind of fading-out definition, like the definition of 'big' or 'old'." ... If, on the other hand, the difference in survival consequences between a putative replicator and its alleles is almost negligible, the replicators under discussion would have to be quite small if the difference in their survival values is to make itself felt. This is the rationale behind Williams's (1966, 25) definition: "In evolutionary theory, a gene could be defined as any hereditary information for which there is a favorable or unfavorable selection bias equal to several or many times its rate of endogenous change." (1982, 89)

Here, Dawkins is not talking about replicators in the general sense of the earlier definitions, but specifically about the "active germ-line" replicator. Whether something functions as a replicator in this sense, and thus whether we are to say that it is a replicator, depends not only on probable longevity in generation time, but also on the specific selection pressures that bear on it as well.

There is one last criteria that Dawkins discusses for replicators. This is what he most cogently calls "the test of mutilation" (1982, 108). Again, the discussion is in the context of what can function as replicators in biological evolution.

> To regard an organism as a replicator, even an asexual organism like a female stick insect, is tantamount to a violation of the "central dogma" of the non-inheritance of acquired characteristics. A stick insect looks like a replicator, in that we may lay out a sequence consisting of daughter, granddaughter, great-granddaughter, etc., in which each appears to be a replica of the preceding one in the series. But suppose a flaw or blemish appears somewhere in the chain, say a stick insect is unfortunate enough to lose a leg. The blemish may last for the whole of her lifetime, but it is not passed on to the next link in the chain. Errors that affect stick insects but not their genes are not perpetuated. Now lay out a parallel series consisting of daughter's genome, granddaughter's genome, great-granddaughter's genome, etc. If a blemish appears somewhere along *this* series it will be passed on to all subsequent links in the chain. It may also be reflected in the bodies of all subsequent links in the chain, because in each generation there are causal arrows leading from genes to body. But there is no causal arrow leading from body to genes. No part of the stick insect's phenotype is a replicator. Nor is her body as a whole. It is wrong to say that "just as genes can pass on their structure in gene lineages, organisms can pass on their structure in organism lineages." (1982, 97)

This then is the "test of mutilation." For something to be a replicator in biological evolution, errors in copying or mutilation by some accident must be passed on intact to future generations. Obviously, without this feature, adaptive variations will not be preserved, even if they do in the first generation increase the reproductive success of the organism carrying the variant. Dawkins builds this necessary feature of Darwinian progress into his criteria for what can be a replicator.

Dawkins's purpose in his characterization of the active germ-line replicator is to specify an entity that is sufficient to create a Darwinian process. That is, it is the entity whose proliferation is governed by the nature of the environment and whose differential proliferation in response to varying environments results in the kind of complex "Paley's watch" adaptations we see in the biological world. The inherent interest in these processes accounts for much of the attraction of Dawkins's replicator. Our difficulty is that this characterization is couched in terms of a system (genetics) in which it is assumed that there are such entities and the question is which entities are indeed the centrally important replicators. We, on the other hand, are interested in determining whether there are replicators in a given process (e.g., cultural evolution) and whether we may be seeing Darwinian effects there.

Replicator Theories

For our purposes, the notion of a replicator (ones that can function in Darwinian processes) that emerges is something like this. The paradigm of a replicator is DNA or, more generally, some sort of self-replicating molecule. More specifically, a particular replicator is conceived of as a type of molecule, with a specific chemical structure, whose tokens are related by descent through the copying process. To function in a Darwinian process, the replicator must be "unitary" enough; that is, the tokens must interact consistently enough with the environment to respond consistently to the selection pressures determined by the environment. Types of molecules have this kind of unitary nature because of the discrete nature of the elements and the chemical bonds of which they are composed. Also, the rate of change of the type (variation) instantiated by a lineage has to be low enough to ensure responsiveness to selection pressures. This requires accuracy in copying, and this is achieved because the molecules act like "molds or templates" for copies of themselves. The characterization of replicators is not intended to be specific to nucleic acids, however, or even to types of molecules. Rather, close analogues of these chemical replicators can function in Darwinian processes as well. Lastly, replicators must pass the "test of mutilation," errors in their generation or accidental damage must be passed on to future generations.

This chapter is not specifically concerned with the genetic replicators with which Dawkins is concerned, but it is important to be as clear as possible about what kinds of things replicators are before venturing to discuss the main question of this chapter: whether there are cultural replicators. It is to this that we now turn.

Memes: The Cultural Replicators

Dawkins believes that DNA molecules are not necessarily the only active germ-line replicators, even though they may be the only ones involved in biological evolution per se. His suggestion for another member of this special class of entities is as follows:

> I think that a new kind of replicator has recently emerged on this very planet. It is staring us in the face. It is still in its infancy, still drifting clumsily about in its primeval soup, but already it is achieving evolutionary change at a rate that leaves the old gene panting far behind.
>
> The new soup is the soup of human culture. We need a name for the new replicator, a noun that conveys the idea of a unit of cultural transmission, or a unit of *imitation.... mimene... meme....*
>
> Examples of memes are tunes, ideas, catch-phrases, clothes fashions, ways of making pots or of building arches. Just as genes propagate themselves in

the gene pool by leaping from body to body via sperms and eggs, so memes propagate themselves in the meme pool by leaping from brain to brain via a process which, in the broad sense, can be called imitation. (1976, 192)

So here we see the idea in which we are interested, the idea that cultural evolution may be a Darwinian process. For Dawkins, this means that there must be cultural replicators, or "memes" – discrete units of cultural transmission or "imitation" that are differentially replicated within culture. Yet, here, Dawkins's claims are much more difficult to swallow than before, and this is specifically because he has set such high standards (and rightly so) for what can function as the replicator in biological evolution. For us to accept the idea of "memes," we need to accept that there are at the center of cultural evolution entities which are enough like DNA, and little enough like organisms or populations, to deserve the name "replicator." He is quick to acknowledge that the analogy may be a bit strained:

> This looks unlike the particulate, all-or-nothing quality of gene transmission. It looks as though meme transmission is subject to continuous mutation, and also to blending.
>
> It is possible that this appearance of non-particulateness is illusory, and that the analogy with genes does not break down.... So far I have talked as though it was obvious what a single meme-unit consisted of. But of course it is far from obvious....
>
> I appeal to the same verbal trick as I used in Chapter 3. There I divided "gene complex" into large and small genetic units, and units within units. The "gene" was defined, not in a rigid all-or-none way, but as a unit of convenience, a length of chromosome with just sufficient copying-fidelity to serve as a viable unit of natural selection....
>
> Similarly, when we say that all biologists nowadays believe in Darwin's theory, we do not mean that every biologist has, graven in his brain, an identical copy of the exact words.... Yet, in spite of all this, there is something, some essence of Darwinism, which is present in the head of every individual who understands the theory. If this were not so, then almost any statement about two people agreeing with each other would be meaningless. An "idea-meme" might be defined as an entity that is capable of being transmitted from one brain to another. The meme of Darwin's theory is therefore that essential basis of the idea which is held in common by all brains that understand the theory. The *differences* in the ways that people represent the theory are then, by definition, not part of the meme. (1976, 195–6)

Fair enough. What the actual units of cultural transmission are really should be an empirical matter rather than being something to be decided by a priori definitions. As such, an approach in which we can proceed with theorizing and defer the nasty empirical questions seems entirely appropriate. But this is not enough to address the question of discreteness. I noted earlier that one reason DNA functions as a germ-line replicator is because of the discreteness of nucleotide sequences. The question as to the size of genes is not the same as the question of what kind of thing they are, and the same is true of memes. It was clear in the case of genes that they were some segment of DNA, and only the question of which segment was at issue; it is the only issue that is addressed by the "verbal trick" just previously appealed to. We are justified in assuming a genetic code because we know a great deal about the physicochemical basis of genetic transmission. We have no equivalent knowledge of the physical basis of cultural transmission and, as such, we are in the position of Darwinian theory before its synthesis with Mendelian genetics. We cannot assume any discrete code for the units of cultural transmission until we have good reason to do so.

The other item of concern here is a kind of systematic ambiguity as to whether the memes are things we see and hear (tunes, ways of making pots) or some sort of (e.g.) brain-state associated with them. Dawkins addresses this difficulty in *The Extended Phenotype:*

> A meme should be regarded as a unit of information in a brain (Cloak's "i-culture"). It has a definite structure, realized in whatever physical medium the brain uses for storing information.... This is to distinguish it from its phenotypic effects, which are its consequences in the outside world (Cloak's "m-culture.")...The phenotypic effects of a meme may be in the form of words, music.... They may be perceived by the sense organs of other individuals, and they may so imprint themselves on the brains of the receiving individuals that a copy (not necessarily exact) of the original meme is graven in the receiving brain. The new copy of the meme is then in a position to broadcast its phenotypic effects, with the result that further copies of itself may be made in yet other brains. (1982, 109)

Now this resolves the ambiguity about where the memes are supposed to be. They are inside the brains, they are the replicators. The things we see people do as a result of being in possession of a certain meme are the "phenotypic" effects of the meme.[4] This makes a certain amount of sense, since in biological evolution the genotype is hidden from casual observation, and it is some developmental consequence of the gene that interacts with the environment and affects rates of proliferation. Likewise, we don't see (or hear) the brain-state that codes for the tune but the tune itself, and we remember the tune or forget it based on that outward, phenotypic effect. This utilization of the rather obvious parallel with the genotype-phenotype structure of biological organisms, while resolving the ambiguity, leaves us with a couple of pressing difficulties and introduces a third.

One of the difficulties that remains unresolved is the one mentioned earlier with the discreteness of the "mimetic" code. We have no reason to assume that ideas are stored in anything like isomorphic sequences of discrete nucleotides, and it would be sheer folly to generate a theory of cultural evolution based on such an assumption, especially if it can be avoided. A related difficulty, which was not mentioned earlier, is the question of whether it is reasonable to assume that there is anything physically similar between the features of my brain that constitute the "essence" of my understanding of Darwinism and the features of your brain that constitute the corresponding essence for you. Presumably, brains are idiosyncratic about how they store ideas, and our theory should allow for this. The result is that Dawkins is in the position of claiming that memes are entities that may bear little resemblance to each other, other than their phenotypic effects, which, while similar, are not identical. These memes seem even less like good candidates for replicator status than populations or even organisms.

The other difficulty that was introduced by the genotype-phenotype model is this: If memes are acquired through the observation of phenotypic effects, then modifications in the phenotypes are inherited, so that the phenotypic effects would seem to be part of the replicator, the meme. The whole point of the phenotype-genotype distinction, however, is that phenotypes are not copied, but here they are. So just what is the replicator supposed to be? Is it the alleged phenotype, the behavior? How can something as ephemeral, as infinitely variable as a behavior, possibly meet the restrictions that Dawkins gives for replicators? This confusion is uncharacteristic of Dawkins, at least in his treatment of biological evolution. What is going on?

Dawkins is apologetic. He recognizes that there are problems with the meme idea. He writes,

> Presumably, as in the case of genes, we can strictly only talk about phenotypic effects in terms of differences, even if we just mean the difference between the behavior produced by a brain containing the meme and that of a brain not containing it. The copying process is probably much less precise than in the case of genes: there may be a certain "mutational" element in every copying event.... Memes may partially blend with each other in a way that genes do not. New "mutations" may be "directed" rather than random with respect to evolutionary trends. The equivalent of Weismannism is less rigid for memes than for genes: there may be "Lamarckian" causal arrows leading from phenotype to replicator, as well as the other way around. These differences may

> prove sufficient to render the analogy with genetic natural selection worthless or even positively misleading. My own feeling is that its main value may lie not so much in helping us to understand human culture as in sharpening our perception of genetic natural selection. This is the only reason I am presumptuous enough to discuss it, for I do not know enough about the existing literature on human culture to make an authoritative contribution to it. (1982, 112)

The point then was not to contribute to the theory of cultural evolution, but to give people another image to think about, an image of an alternative to DNA as the only replicator. There is no real reason to correct the difficulties noted earlier or to retract the suggestion of "mimetic" Darwinian processes, since it was only intended for purposes of illustration. A similar comment occurs in the notes to the "new" edition (1989) of *The Selfish Gene:*

> ...my designs on human culture were modest almost to the vanishing point. My true ambitions – and they are admittedly large – lead in another direction entirely. I want to claim almost limitless power for slightly inaccurate self-replicating entities, once they arise anywhere in the universe.... The first ten chapters of *The Selfish Gene* had concentrated exclusively on one kind of replicator, the gene. In discussing memes in the final chapter I was trying to make the case for replicators in general, and to show that genes were not the only members of that important class. Whether the milieu of human culture really does have what it takes to get a form of Darwinism going, I am not sure.... My purpose was to cut the gene down to size, rather than to sculpt a grand theory of human culture. (1989, 322f)[5]

Well, Dawkins is after all a biologist, and he never asked anyone to take his suggestions regarding cultural evolution seriously. Still, as he says earlier in that note to the "new" edition, "The word *meme* is turning out to be a good meme. It is now quite widely used and in 1988 it joined the official list of words being considered for future editions of Oxford English Dictionaries" (1989, 322).

Those of us interested in cultural evolution may find ourselves using the word *meme*, usually in conversation, when we want to talk about a unit of cultural transmission without having to specify if it is an idea or a tune or a behavior and without worrying how big those things are or what they are made of. We need such a word, and Dawkins, in typical style, has given us one. But concepts are not just names. They bring with them clusters of assumptions. In this case, the assumption that the units of cultural transmission are identifiable with brain states and are a matter of some sort of similarity between different people's brain-state; that they "leap from brain to brain via imitation just as genes leap from body to body via sperms and eggs." Most problematic is the assumption that memes are replicators, that for there to be units of cultural evolution there has to be something that is an awful lot like DNA.

The point here is not to blame Dawkins for the misapprehension that a lot of people are operating under, the assumption that cultural evolution needs replicators. The point is not even to present a diagnosis as to why this assumption is so prevalent. The point is to demonstrate the nature of this assumption. Even in his disclaimers, Dawkins manages to set a research agenda. As a recognized authority on evolutionary processes, he says that without replicators of the kind that he is interested in, he sees no way to get cultural Darwinian processes going. The challenge, if we accept it, is to find such replicators. In the remainder of Parts I and II, I argue that the right response is to deny this challenge, to create models of Darwinian processes without replicators, ones which are appropriate for cultural evolution and consistent with what we know about the mechanisms of thought and behavior. But first, let us examine a few attempts to take up this challenge. We begin with David Hull's *Science as a Process*.

DAVID HULL: REPLICATORS AND THE EVOLUTION OF SCIENCE

David Hull's (1988a) *Science as a Process* stands as the most ambitious attempt (to date) to flesh out the details of the evolutionary theory of science. The work (the 522-page length of which has elicited comment from a number of reviewers) is divided roughly into two parts. The first is a history of the development of evolutionary theory and systematics (biological taxonomy), with the emphasis on recent power struggles between various camps in the systematics community. Of particular interest is the detailed investigation of editorial practices in the journal *Systematic Zoology*. The upshot of his analysis is that the vicious polemics and accusations of editorial abuse which have characterized the systematics community for the last two decades are not as severe as critics of science would have them be, and the residual infighting somehow manages to work toward the progress of science in the longer term. It is not this aspect of Hull's work that causes it to find a place in this chapter, but the second half of the book in which he develops a general model of selection processes and applies it to science. More particularly, it is the fact that his general model is a replicator model that concerns us here.

Hull agrees with Dawkins that replicators are needed for Darwinian evolution; indeed, as we shall see, he builds replicators into his definition of selection. He does not, however, adopt Dawkins's approach to doing evolutionary analysis whole. He diverges from Dawkins on two main points. First, he faults Dawkins for paying insufficient attention to the phenotype and to environmental interaction. Dawkins's model tends to place all of the emphasis on the gene, and phenotypes are relegated to the role of "vehicles" of the gene, an emphasis that was only partly corrected in *The Extended Phenotype*. This is a general sort of objection that many people have to gene selectionism and to the population genetics approach to doing evolutionary theory. The second main difference is more important. Hull believes that the whole "units of selection" debate and the attempt to find the level at which selection occurs are misguided. Instead, he advocates a hierarchical approach in which interaction resulting in selection can occur at a variety of levels, leaving it as an empirical matter on which level interaction is actually occurring. The "levels" concerned in biological evolution are allele, gene, genome, organism, group, species, genus, and so forth. What the corresponding levels are in cultural evolution is an open question, but the point is to build a general model which can accommodate selection on a variety of levels, whatever they turn out to be.[6] As such, Hull's definitions of the central components of Darwinian processes need to be more explicit than Dawkins's. Particularly, to facilitate multilevel analysis as well as to emphasize the importance of the site of interaction with the environment, the *interactor* is introduced as a central component along with the *replicator*. Selection is defined in terms of a causal relationship between replicators and interactors, and the *lineage* completes the catalog of central components of the general model.

The central definitions are as follows:

> *replicator* – an entity that passes on its structure largely intact in successive replications.
> *interactor* – an entity that interacts as a whole with its environment in such a way that this interaction *causes* replication to be differential.
> *selection* – a process in which the differential extinction and proliferation of interactors *causes* the differential perpetuation of the relevant replicators.
> *lineage* – an entity that persists indefinitely through time either in the same or an altered state as a result of replication. (Hull 1988a, 408f)

For Hull, the notion of a "lineage" is central to the general model of selection processes, perhaps because of its importance in the modern understanding of species. Roughly speaking, what makes two replicators the same for Hull's purposes is that they are related by descent, that they are part of the same lineage. This stands in contrast to Dawkins's account, in which two strands of DNA contain the same gene just in case they contain segments with the same chemical structure. There, the emphasis was on identity of structure; here, it is on relatedness by descent.[7]

The central components of selection processes Hull defines are not abstract entities, on his account. On the contrary,

> By now, one thing should be clear: everything involved in selection processes and everything that results from selection are spatiotemporal particulars – individuals. Both replicators and interactors are unproblematic individuals. To perform the functions that they do they must have finite durations. They must come into and pass out of existence. (Hull 1988a, 411)

Replicators have to be spatiotemporal particulars, then, as are everything else that will appear in an application of Hull's general model. The advantage of this approach is that it sets the stage for a materialistic account of cultural evolution. We will not be sidetracked with difficult philosophical questions about universals and "meanings" – just the kind of thing that Hull wants to avoid (1988a, 29f). The difficulty with this is that spatiotemporal particulars cannot be in more than one place at a time. If the replicators in science are (e.g.) ideas, then we will want to talk about the dissemination and proliferation of particular ideas and about the number of people that have the same idea. It is not so much the spatiotemporal particular (the token) but the type that interests us. This difficulty aside, the thing to remember here is that Hull uses "replicator" to refer to the token, rather than to the type, and his emphasis is on the lineage rather than the type. The relevant type is derived from the "structure" that is passed on (largely intact) from one replicator to the next. Types of replicators, as defined by common structure, however, may be relatively ephemeral in cultural evolution, as opposed to having the kind of virtual "immortality" of Dawkins's biological replicators. The lineage, on the other hand, is determined not by common structure but by the chain of descent, down which unlimited change is possible. It is the ultimate survival of the lineage that Hull's definitions would have us watch, rather than the survival of the particular replicator type – say, a particular idea or belief (Hull 1988, 411).[8]

What *interactors* interact with is the environment. Hull's reason for emphasizing the role of interactors is that it is at the level of interaction that the discriminating aspect of selection occurs. Replicators are not selected directly according to their characteristics, but according to their consequences. If some replicator were differentially copied because of its immediate physical characteristics, it would be both replicator and interactor. (Hull mentions this possibility on p. 410.) An example of this would be Dawkins's primordial replicators, self-replicating molecules floating loose in the chemical-rich soup of the oceans. Such replicators function as their own interactors, rather than generating some other protecting and supporting structure, whose efficacy determines the replicative success of the replicators. This can be thought of as a degenerate case of the replicator-interactor relationship, in which that relationship is reduced to mere identity.

For Hull, the relation between replicators and interactors is clearly supposed to be similar to that between the genotype and phenotype in biological evolution. The success of the interactor affects the replication of the associated replicator. Hull does not want to say that the replicator is the *cause* of the interactor (as we are prone to say in the case of biological evolution) since this would create problems for the claim he wants to make that scientists are the primary interactors in science. Presumably, however, there must be some sort of mechanism to ensure that particular kinds of interactors are associated with particular kinds of replicators – or better, that particular kinds of interactions result from particular kinds of replicators.

There is a certain counterintuitiveness about the way Hull defines interactors. From the standpoint of the construction of a general model, one expects that we should be interested in talking about the particular consequence of the replicator, but here one interactor can function as the interactor for a number of replicators. With the traditional genotype-phenotype distinction, we see the organism as the relevant phenotypic component for the entire genome, and the trait as the relevant phenotypic component for the particular gene. But Hull's replicators are more like individual genes, and his interactors, rather than being like traits, are like organisms. The reason for this will become clearer when we see how he applies the model to science, but first a couple of things need to be said about the last central component in his model: selection.

Hull's intent is that *selection* be defined simply in terms of the definitions given of replicators and interactors (Hull 1992, 235). Selection is "a process in which the differential extinction and proliferation of interactors *cause* the differential perpetuation of the relevant replicators" (1988a, 409). A simple causal relation between what happens to interactors and the perpetuation of replicators completes the specification of selection. What he means here is clear enough, but two things should be kept in mind. First, this is supposed to be a general model of selection processes, applicable to biological as well as cultural evolution. Thus, Hull's selection processes are equivalent to what I have been calling "Darwinian processes," those processes in which we can get the interesting kinds of effects that we see in biological evolution by natural selection. As such, Hull's definition includes the implicit claim that replicators are necessary for such processes. Hull also states repeatedly that descent through replication is necessary for selection processes,[9] so there can be little doubt that his position on this point is fundamentally at odds with the thesis being pursued in this book – that replicators are not necessary for such processes.

Second, notice that the role of interactors is defined in such a way that extinction and proliferation of the interactor is not necessary for the differential proliferation of replicators. This allows that an interactor's failure can cause the extinction of a line of replicators without requiring the elimination of the interactor itself, and if we are to think of scientists as the interactors for their ideas or theories (the replicators), then this is as it should be. The difficulty is that *selection* is defined so that extinction of the interactor *is* necessary, which implies, at the very least, that the failure of the proliferation of an idea requires the loss of professional status of the scientists holding that idea. While this kind of thing may occur, even with some regularity, there is good reason to think that ideas may frequently lose currency even though no individual's professional standing is lost.[10]

The impression resulting from the examination of Hull's general model of selection processes is that its real purpose is not to be a general model, but rather a special model for a particular kind of analysis, despite his claims of its generality. That this is the case we will see by examining Hull's application of the model to scientific evolution.

Science as a Selection Process

So what are the *replicators* in science? Hull writes,

> The answer is not very surprising: elements of the substantive content of science – beliefs about the goals of science, proper ways to go about realizing these goals, problems and their possible solutions, modes of representation, accumulated data, and so on. Scientists in conversations, publications, and classroom lectures broach all of these topics. These are the entities that get passed on in replication sequences. Included among the "vehicles" of transmission in conceptual replication are books, journals, computers, and of course human brains. (Hull 1988a, 434)

Clearly, "replicator" is supposed to cover a wide variety of entities in science, including elements of practice and linguistic habits ("modes of representation"). But, as is evident in this passage, Hull moves quickly to concentrate on "conceptual" replication, by which he means not only the basic concepts which form the vocabulary in which scientific beliefs and theories are expressed, but those beliefs and theories as well (Hull 1988b). The initial intimation that he is going to discuss mechanisms for the proliferation of practices, perhaps in the absence of linguistic "vehicles," is never justified. Instead, his discussion proceeds as though practices formed part of the environment, or part of the "interactor," which are somehow coded for by the informational content of the linguistically characterized replicators – something of a disappointment for those of us desiring a model covering transmission of behaviors which is not linguistically mediated.

The *interactors* in science are as follows:

> Conceptual replicators cannot interact directly with that portion of the natural world to which they ostensibly refer. Instead, they interact only indirectly through scientists. The ideas that these scientists hold do not produce these scientists in the way that genes produce organisms, but they do influence how they behave. Scientists are the ones who notice problems, think up possible solutions, and attempt to test them. They are the primary interactors in the conceptual development of science. Hence, in conceptual change, agents such as scientists function as "vehicles" in both Campbell's sense and Dawkins' sense. (1988a, 434)

Campbell's (1974) sense of "vehicle" is the sense in which a particular strand of DNA is a vehicle for the genes that are instantiated there: the vehicle is the carrier for the structure or information that defines the replicator-type (here using "replicator" in Dawkins's sense of the replicator-type rather than Hull's replicator-token). Dawkins's sense of "vehicle" is the phenotype – or what Hull terms "interactor," the site of the all important causal interaction at which the environment "makes the choice" of how replicators will be propagated. (Note that Hull's understanding of the replicator-interactor relationship is not necessarily developmental.) Given that this is supposed to be a general model, one wonders why the multiple roles that scientists play in the logic of the selection process are not clearly distinguished in the basic definitions of the central components. The answer lies in Hull's account of the particular dynamics of scientific evolution.

The most interesting and substantive things that Hull has to say about science have little to do with the general model of selection processes outlined earlier. The concepts his argument turns on are not replicator, interactor, and selection, nor does he worry much about the adequacy of his definitions of these terms for capturing the full dynamics of scientific evolution. Instead, he focuses on other notions: credit, curiosity, checking; competition and cooperation; cheating and stealing; and, most important, conceptual inclusive fitness. Scientists act so as to maximize their "conceptual inclusive fitness." This means that they act so as to maximize the proliferation of their ideas. The substance of Hull's account of science lies in his explanations as to (1) why they do this, and (2) how the scientific institutions constrain this drive in such a way that the results are good for science in general, so that the ideals of science are systematically realized.

Hull takes for granted that scientists are curious and that they desire credit for their contributions. The former assumption is warranted because curiosity is a prerequisite for being a scientist, the latter because it is in their nature as status-seeking social animals. The desire for credit drives the attempt to maximize conceptual inclusive fitness: "Increasing one's conceptual inclusive fitness in science means increasing the number of replicates of one's contributions in the work of successive generations of other scientists" (1988a, 283). "Inclusive," as it is used here, deserves a bit of explanation.

The term *inclusive fitness* was introduced by W. D. Hamilton in his 1964 article, "The Genetic Evolution of Social Behavior." The reason for its introduction was the realization that selective forces operate on morphological and behavioral traits not only according to their effect on the reproductive success of the individual carrying the trait, but also according to their effects on the reproductive success of other individuals carrying the trait. Thus, the inclusive fitness of a gene is calculated according to its effects on the proliferation of gene-tokens of its own type, whether or not it is itself a member of the lineage affected, and the process that selects for positive contributions to the reproductive success other than descendants is called "kin-selection." The primary application of the notion of kin-selection has been to the explanation of apparently "altruistic" behavior, and this was also Hamilton's concern in his 1964 article.

There is a predictive thesis that goes along with the notion of conceptual inclusive fitness: "Scientists should behave in ways to increase their conceptual inclusive fitness, energy flow should follow the flow of ideas" (1988a, 287). The "shoulds" here are not normative but are intended to be predictive of the behavior of scientists, given their nature as credit-seeking social animals. The obvious parallel is with biological evolution, in which we can expect the energy flow to follow the flow of genes. Likewise, scientists "invest" in what is most likely to maximize their influence and bolster their reputation. But this is not the end of the story. Just as genes are not fabricated from the ground up at each mutation or recombination (as some of the religious antievolutionists have supposed in their calculations), neither are scientific theories. Scientists inherit most of the substance of their theories from other scientists, and the quality of contributions from predecessors is crucial to the success of the new contribution. Thus, rigorous "checking" is also driven by conceptual inclusive fitness. This checking is the equivalent of selection, so that rigorous checking is like intense selection pressure, and the drive for conceptual inclusive fitness ensures that checking will occur both in the evaluation of earlier contributions which are to be incorporated and in original contributions. For one knows that other scientists are just as concerned with their conceptual fitness, so that if your contribution fails to pass their self-interested checking, it will not be further propagated, thus reducing your conceptual fitness.

Thus, from the assumption that scientists are credit-seekers in an environment in which the main measure of success is the number of replicates of your ideas that there are, we get the rigorous criticism of the ideas of others and great care in what is published. Conceptual inclusive fitness ensures that the institutional norms of science are obeyed; selfish scientists can be just as good at contributing to the progress of science as selfish genes are at making evolution happen. This is the main thrust of Hull's answer to the antiscience types, but he gets a bit more explanatory mileage out of it than just these points.

Hull notes that cheating (fraudulent research) is much more severely punished in science than is stealing (claiming credit for the contributions of others). The reason for this is, again, the drive to maximize conceptual inclusive fitness. It really doesn't matter where ideas come from if they are good. The only person hurt by stealing credit is the person from whom it is stolen, and perhaps the thief if he is caught. Fraudulent research generally results in bad ideas, however, and hurts everyone who might possibly use the ideas. Thus, the conceptual inclusive fitness of large numbers of scientists is threatened by fraud, whereas the number threatened by theft is relatively small. Therefore, the punishment for fraud is much more serious, usually resulting in the loss of professional standing.

So far this all sounds as though scientists work in isolation, which is clearly not the case, nor is it Hull's view. Scientists form groups in a variety of sizes, from the research team, to "demes" – consisting of dozens to hundreds of scientists, up to "invisible colleges." The reason for the research teams is fairly obvious: Division of labor allows specialization, allowing more sophisticated or time-consuming projects, which maximize the influence of all members of the teams. The demic group structure in science is more interesting.

In biology, a "deme" is a portion of a gene pool within which there is higher than average exchange of genetic material, the members of which have lower than average exchange with those outside the deme. By analogy, in science a deme is a group of scientists who use each others ideas and recognize each other's contributions more frequently than those of scientists outside the group.

One important (easily verifiable) aspect of scientific demes is citation patterns. The scientist wishes not only that her ideas get propagated but that she gets credit for them as well. One of the ways that the recognition of contributions is expressed is in citations. Hull suggests that citation patterns both indicate and reinforce deme structure.

> At the very least, positive citations indicate which work a scientist thinks lends greatest support to his or her own research. Scientists give credit not so much where credit is due, but where it is useful. In a word, science is a function of *conceptual inclusive fitness*. Scientists behave in ways calculated to encourage other scientists to use their work, preferably with ample acknowledgment of that use. Conversely, the best thing that one scientist can do for another is to use his or her work, preferably with ample acknowledgment. (1988a, 310)

Now scientists, unlike biological organisms, can serve as reciprocal conceptual "predecessors" for each other so that, if citing someone is a favor, it is a favor that can be returned. What these sorts of considerations result in is patterns of mutual citation, which enforce the demic structure of science. This is not to suggest that scientists do not cite one another because they are working on the same subject or that they consider each other's contributions important. The point is that these obvious considerations of worth are not the only factors that sustain demes. These other factors, this "I'll cite you if you cite me," is just the kind of thing that critics of science like to bring up; it seems to be a nonfunctional behavior aimed solely at the selfish ends of scientists – in this case, groups of them. Hull's purpose, again, is to show how apparently "improper" behavior on the part of scientists actually furthers the appropriate ends. For this, we need to understand the effects of demes in biological evolution.

In the most common models of population genetics – for instance those in which Fisher's (1930) "fundamental theorem" is expressed – it is assumed that every individual member of a species has an equal probability of mating with any other member of the population, and the population size is effectively infinite, eliminating the possibility of "drift" caused by sampling error. Sewall Wright's (1932, 1986) "shifting-balance" theory of evolution, on the other hand, capitalizes on just this phenomenon, and this is where Hull gets his ideas about the effects of demic structure on evolution.

One consequence of Fisher's fundamental theorem is that, in effectively infinite populations with random mating, the mean fitness of the genes in the population's gene pool will increase to a maximum. This means that genes with lower than average fitness will be eliminated, unless their fitness is increased as they become more rare. This is not necessarily true in small populations, however. Here, sampling error can cause average fitness to decrease, which means that the frequency of some less fit genes must increase.

Wright's "shifting-balance" theory of evolution supposes that this genetic drift is central to the process of evolution. In real populations, there are "demes," which effectively form genetic subpopulations of limited size in which drift is pronounced. Demes become the scene of rapid, although usually disastrous evolution caused by drift, and selection then occurs between demes. This introduces a new central evolutionary force to the theory of evolution, and a central role for demes – subpopulations of limited size.

Hull writes, "One contention of the present work is that the small research groups that periodically crop up are the most important focus of rapid, though usually abortive change in science. They are the locus of initial innovation and evaluation" (Hull 1988a, 112).

As to what the mechanisms are that account for this, Hull speculates, "One would expect that presence of research groups and conceptual demes to have the same effects on conceptual evolution that population biologists have shown that kinship groups and biological demes have on biological evolution" (Hull 1988a, 514).

What Hull is referring to here is just the role that demes play in Wright's shifting-balance theory. (He does not develop the implicit parallel between kin-group altruism and the internal behavior of research groups.) Thus, the primary explanation that Hull offers for the rapidity of conceptual change within the smaller scientific groups is by parallel with that occurring within biological demes. The impression that this gives is that he intends that rapid conceptual change within research groups is to be explained by some sort of drift phenomena. This does not turn out to be the case.

The mechanisms that Hull actually suggests are as follows:

> One function of the smaller groups is to provide sympathetic criticism while a scientist develops his or her ideas. (Hull 1988a, 366)
>
> One also increases the chances of having one's work noticed by becoming part of one or more informal groups of scientists....Members of these groups also come to each other's aid after publication. Not only can they keep the ideas of their allies from being ignored, but also they can guarantee that at least a few of the published responses will be positive. (367)

These suggestions have the flavor of observations about what scientists actually do, and this is where Hull finds himself on firm ground because of his extensive research. Part of the reason that small groups in science are the scene of so much rapid change is that the atmosphere is more cooperative and supportive, criticism less energetic. The difficulty is that the appropriate parallels to biological evolution here are increased fecundity and decreased intensity of selection. Yet, nowhere does Hull discuss biological phenomena in which fecundity is locally increased or where the small size of a population results in decreased selection pressure for its members. Nor does he discuss what the parallel of genetic drift would be in conceptual evolution, other than the suggestions that there must be parallel mechanisms of some sort.

What seems to be the case is that Hull did not seriously intend that parallels to Wright's shifting-balance evolution account for all of the rapidity of change in small groups. Rather, there seems to be a variety of mechanisms involved, including some that may have no parallel in biological evolution, and Hull would probably agree with this. These worries only bear peripherally on his project, however. The only essential point was to establish that small groups are an essential source of novelty in science, and thus the behaviors which reinforce the structure of these groups, as "improper" as they may seem to critics of science, actually contribute to the progress of science. Whether something such as "conceptual drift" is responsible for the phenomenon is of secondary importance.

The biggest difficulty with assessing Hull's general model of selection processes stems from the lack of polish in the details of the model itself. Given that his primary purpose is not, after all, the construction of the general model but rather the presentation of an in-depth case study of conceptual change and the behavior of the scientists involved addressed at "externalist" critics of science, it is perhaps uncharitable to complain that he has not filled in these details to our satisfaction (even though he presents his model as one adequate for the general analysis of selection processes). On the other hand, it is not our task here to finish what Hull started, since our thesis is that he has chosen the wrong starting point. As such, I restrict my criticism to issues bearing on the general tenability of replicator models of the type that Hull proposes and to pointing out weaknesses of Hull's model as it stands.

The strength of Hull's model is that it allows one to say things about the behavior of scientists and the structure of scientific groups of various sizes in a fairly succinct way. Ideas are the replicators which pass on their structure more or less intact; scientists are the interactors whose successes or failures determine the rate at which ideas are reproduced. Good ideas make for successful scientists, bad ideas make for failures. Scientists are dedicated to the propagation of their ideas, their "conceptual inclusive fitness." This leads to energetic attempts to explain the world correctly and to rigorous checking of those attempts. The supposedly "improper" behavior of scientists can be better understood in this light. Behaviors that affect group cohesion increase the rate at which conceptual change occurs; intergroup "warfare" intensifies selection pressures.

If one wants a model that lends itself both to formalization and to convenient and systematic application, Hull's leaves much to be desired. His "general model" does go beyond Dawkins's emphasis on replicators with the addition of interactors and the general definition of selection, but apart from that it is encumbered by simplistic analogies with biological processes that do little to illuminate the underlying logic of Darwinian processes and shows little thought for the special difficulties of modeling cultural evolution. Ideas are the replicators (at least the only ones that Hull discusses), but what are ideas? Presumably, they must be spatiotemporal particulars, and while this departs somewhat from common usage, we should be prepared to allow this to a materialist theory. We are to focus on tokens and lineages of tokens rather than types. But what are the tokens of ideas? Common sense is no real guide to this question since it tends to yield dualism. Ideas could be brain-states or behaviors (or perhaps behavioral dispositions). At this point, we should remember the discussion of Dawkins's meme theory. It makes no sense to suppose that having the same idea means having the same brain-state. On the other hand, behaviors are not spatiotemporal particulars in the sense that one can speak of someone "having" a behavior in the same way we speak of them "having" an idea. Dispositions seem like the most likely candidate (which I explore in Chapter 2) but Hull does not discuss them, and indeed, as replicators they suffer from the same deficiency as brain-states. Hull's model, like Dawkins's, seems acceptable just so long as we assume the tenability of some sort of structurally identical token proliferated through imitation or communication, but answers to the question as to how this is supposed to work are not forthcoming.

Discussion of the transmission of "ways" of doing things never occurs, nor is the mechanism of transmission of ideas ever made clear. Of course, it is a good idea to leave this kind of thing up to the specialists, but in doing so we should leave ourselves as open as possible to whatever they may come up with. Rosenberg quips that Hull's theory "will be held hostage to the fortunes of propositional attitude psychology" (1992, 224), which Hull accepts: "If the world turns out not to be the way my theory requires, then my theory is false" (1992, 233f). Just what does his theory require? That there is some spatiotemporal particular that is passing on its structure more-or-less intact when an idea is passed on, and the process of this passing can sensibly be called "replication," by analogy to the replication of DNA molecules. If being the "same" idea is fundamentally a matter of "family resemblance" as Wittgensteinians keep insisting, then Hull is out of luck. Even if things do turn out to be right for Hull's theory, the point remains that his replicators do not replicate themselves, so that whatever entity is doing the replication becomes central to the process and thus deserves a central place in the model.

The status of scientists as interactors is equally problematic. Given that Hull wants to talk about the success and failure of scientists and of "generations" of scientists, it is convenient that they occupy a central place in his theory, but if they are supposed to be analogous to the phenotype (which seems to be what Hull originally intends by "interactor"), then we are left with a puzzle. The idea seems to be that the scientist qua scientific interactor is determined by the scientific replicators – ideas and so forth – that he carries. Thus, his success or failure depends on the qualities of those replicators, but the puzzle concerns where the interaction is supposed to take place. Is it between the scientist and her environment or within the scientist? The answer seems to be "both." Replication may fail because a scientist rejects the idea or because the scientist loses the ability to pass on her ideas. While this characterization of interactors may allow us to say things about why scientists act the way they do, it gives us no help in talking about what makes some ideas better than others. In the biological theory, we want to know the trait that the gene codes for, so that we can isolate the gene's contribution to the reproductive success of the organism. In Hull's model, there is no component corresponding to the trait. All we can say about the qualities of ideas is that scientists accept or reject them according to whether they think that they will contribute positively or negatively to their success. Why scientists think this is left open.

Hull likes to deny that he is doing *evolutionary* epistemology because he denies he is doing epistemology at all. This protest seems puzzling at first, since one tends to assume that anyone attempting to construct an evolutionary account of science must share (at least) Popper's falsificationist motivations. The fact that Hull places scientists as the interactors in his application of his general model to science, however, demonstrates that his interest in Darwinian processes is to explain how science progresses in terms of the "improper" behavior of scientists, rather that how it progresses in spite of barriers to epistemic access to the world. His account begins with the assumption that scientists somehow know good ideas from bad, and the task of his Darwinian model is to explain how they are motivated to do the work of acquiring better ideas and getting rid of worse ones. The work of the epistemologist is not even begun, nor is that of the selection theorist. Now given that he (correctly) assumes that selection processes are taking place on a variety of levels, then his account of the motivational structures of the scientific community might be compatible with an evolutionary epistemology. His insistence that selection processes must have replicators creates enormous difficulties for the evolutionary epistemologist, however, because of the problems involved in

Replicator Theories

assuming that there are the right kind of spatiotemporal particulars to function as replicator-tokens. On the other hand, what he says about the functioning of scientific communities is not particularly dependent on his general model (as noted earlier) so that his more substantial contributions can most likely be incorporated within a more flexible general model.

DENNETT'S INTENTIONAL-INFORMATIONAL REPLICATORS

Dawkins invented the meme to illustrate the point that genes need not be the only replicators. That is, there is no *intrinsic* property that picks out these segments of DNA other than their durability over time in the population-level shuffling of genetic material. In principle, any "shuffling" evolutionary process must find its largest stable elements, and those elements will be the beneficiaries of natural selection. The basic insight is clear enough, but selling it by *reifying* the replicators themselves at the center of any evolutionary process creates a number of problems. Three have stood out. First is the implication that we cannot apply evolutionary analysis to a system unless there are replicators. Chapter 3 is devoted to explaining what is wrong with that idea. Second, replicators are odd sorts of things. Whether something is a replicator cannot be determined by examining the thing itself, but you must see how it fares in its environment. A replicator is *by definition* a survivor of the shuffling, if not the selective, process. If the environment changes so that selection pressures are lessened, then perhaps what was a replicator is no longer one; frequency change due to endogenous causes overwhelms change due to selection. If this weren't enough, replicators are types whose properties change over time. Dawkins felt the pull of the lineage view, as Hull did, but wanted to retain the structurally defined categories that facilitate scientific observation and measurement. So he has it that genes are individuated by structure but are also the same gene as their ancestors, even if there are mutations. Hull gets the metaphysics right – if something's properties are going to change, it had better be a thing, a metaphysical particular. This rests on a basic principle of methodology. Types whose properties change are fairly useless for scientific purposes, unless they are clearly defined in terms of types whose properties do not change. I don't mean to argue about which kinds of things there *really* are and are not in the world. Rather, the practical requirements of scientific practice and communication put requirements on us that have the force of laws of logic. What science does is mostly a matter of putting things in categories. Change is described by putting things in different categories. The replication and sharing of scientific results requires stable categories. The third problem concerns empirical adequacy: even if we can figure out how to make objective identifications of Dawkins's replicators, there probably won't be enough of them for an interesting account of cultural evolution. There are too many cases of one-shot communications, for instance. For the purposes of an evolutionary epistemology or, say, a theory of consciousness, one needs a way of analyzing the way thought and communication occur *in general* over time.

In the end, Dawkins claimed to have no real ambitions to a general theory of cultural evolution (although he seems willing to accept credit as the meme's architect from those who do have such aspirations; see Blackmore, 1999, for instance). Rather, the current popularity of the meme as a theoretical entity is in large part due to the contributions of philosopher Daniel Dennett. Dennett's meme is, despite his acknowledgment of Dawkins, quite a different breed of cat. In particular, Dennett's meme *is* intended to be broad enough to describe thought and communication in general, not so much because he needs a theory of cultural evolution, but because he needs a way to found the autonomy of the conscious subject within the naturalistic scheme.

Dennett's (1991) construction and use of the meme needs to be understood in the context of his theory of consciousness. To begin with, Dennett is a scientific materialist of sorts, who wishes to give a scientifically respectable theory of consciousness – one that is consistent with our current understanding of human beings as the product of evolution. On the other hand, he shares with a lot of materialists the conviction that, by the time all the details are worked out, the materialist perspective will not have the dire consequences that critics worry about: depriving life of meaning, robbing us of free will, or reducing morality to an unsatisfying emotivism.

On Dennett's account, consciousness is an *emergent* property, but without the mystery that label sometimes brings with it. It emerges in the elaboration of linguistic communication and the simpler signaling systems from which that evolved. Behaviors that evolve as responses to signals sent by others can be triggered by one's own utterances. Such autostimulation loops may manifest themselves in familiar phenomena such as "talking to yourself." External dialogues become internalized. We learn to represent ourselves in order to anticipate other's responses to us, and so forth. The general scheme seems to be an eminently reasonable – indeed, rather obvious – materialist account of the evolution of mind.

Critics will say that all of this may explain the structure of consciousness *in fact,* but it still leaves us as the sort of "lumbering robot" vehicle of the genes about which Dawkins waxed so poetic. The story, no matter how fascinating, well supported, and adequate to the facts, provides only the most deflationary account of those things that matter most to us. It is for this purpose, to answer these worries, that Dennett unveils his meme.

Dennett's contribution to the ongoing evolution of the meme concept occurs principally in *Consciousness Explained* (1991) and *Darwin's Dangerous Idea* (1996). He introduces the idea in a fashion reminiscent of Dawkins.

> Then a few billion years passed, while multicellular life forms explored various nooks and crannies of Design Space until, one fine day, another invasion began, in a single species of multicellular organism, a sort of primate, which had developed a variety of structures and capacities (don't you dare call them preadaptations) that just happened to be particularly well suited for these invaders. It is not surprising that the invaders were well adapted for finding homes in their hosts, since they were themselves created by their hosts, in much the way spiders create webs and birds create nests. In a twinkling – less than a hundred thousand years – these new invaders transformed the apes who were their unwitting hosts into something altogether new: *witting* hosts, who, thanks to their huge stock of newfangled invaders, could imagine the heretofore unimaginable, leaping through Design Space as nothing had ever done before. Following Dawkins (1976) I call the invaders *memes*, and the radically new kind of entity created when a particular sort of animal is properly furnished by – or infested with – memes is what is commonly called a *person*. (1996, 341)

For those who were attracted to Dennett's account of the emergence of consciousness because it seemed so reasonable, this abrupt integration of the scientific fringe may be rather startling. We were talking about the evolution of biological organisms, language, brains, and the like. To be sure, we were also talking about intentional properties, but that particular sort of baggage, however suspect it may be to materialists or eliminativists, is a sort of baggage that we are stuck with and, in any event, Dennett (1987) has an elaborate defense. Memes, however, are a different matter. Moreover, the casual manner in which he handles the meme is not likely to inspire confidence. We return shortly to the examination of the legitimacy of Dennett's meme, but first we need to understand exactly how he seeks to employ it.

On Dennett's account, life emerged via the original replicator, DNA, whose adventures we need not recount here. At some later time, out of the activities of the bodies that colonies of DNA generated to facilitate their reproduction, a second replicator emerges, the meme. At some point the patterns of imitation cross some threshold beyond which a new sort of replicator exists. The replication process involved is just ordinary imitation, the details of which can be integrated once we understand them. The new replicator – the meme – is in some sense a parasite, in that it can propagate simply because of its own propagatability, and in some sense is not a parasite, in that it emerged *from* the system which it parasitizes. It is more like a virus than a living (uni- or multi-cellular) parasite. In Dennett's view, by and large, a happy symbiosis exists between us and the memes, they being so dependent on the choices we make in replicating them.

Like any other meme theorist, Dennett makes free use of the "selfishness" of memes[11] in accounting for the popularity of ideas and institutions which, not to put too fine a point on it, he doesn't like. In general, however, the consequence that he draws from the story of the emergence of memes is not that we are somehow generally at odds with our ideas (cf. Brodie 1996). The twist the story takes in his hands is distinctive. The slogan of the meme's-eye view is "a scholar is a library's way of making another library." The critic, of course, finds this no more reassuring than Dawkins's observation that people are just genes' ways of making more genes. Dennett doesn't think this is going to be a problem. We need not worry about the selfishly replicative tendencies of memes, for

> it cannot be "memes vs. us," because earlier infestations of memes have already played a major role in determining who or what we are. The "independent" mind struggling to protect itself from alien and dangerous memes is a myth. There is a persisting tension between the biological imperatives of our genes on the one hand and the cultural imperatives of our memes on the other, but it would be foolish to "side with" our genes; that would be to commit the most egregious error of pop sociobiology. Besides, as we have already noted, what makes us special is that we, alone among species, can rise above the imperatives of our genes – thanks to the lifting crane of the memes. (1996, 365)

So, for Dennett, what memes do is make us special, make us unique among species, make us free from the unacceptably confining determination of our genes. They are not introduced to explain cultural "progress," nor to explain cultural pathologies, however convenient they may be for that purpose. If they are parasites by their very nature, we need not worry about it, for we *are* the parasites – or, at least, rather than parasitizing us, to a large extent they *compose* us.

However appealing or unappealing, plausible or implausible one finds Dennett's account of the nature of persons, what matters here is what his memes are and for what he uses them. The latter question should be reasonably clear by now. As for what they are, Dennett relies heavily on references to Dawkins, although it becomes quickly apparent that memes in this case are a rather different sort of thing.

Similarities first: Dennett cites Hull in insisting that "we do not want to consider two *identical* cultural items as instances of the same *meme* unless they are related by descent" (1996, 356). Which is to say, there are lineage requirements for the meme. He shares with Dawkins the conviction that one can "finesse" the precise specification of what the meme is in the same way that Williams (1966) finessed the gene. For Dawkins (as for Williams's gene) it was partly a dodge (for not knowing what genes there actually are) and partly a principled formula for identifying said genes. In Dennett's formulation, it is pure dodge, because of the lack of any assumed uniform substrate.

> Where evolutionary theory considers information transmitted through genetic channels, whatever they are, cognitive science considers information transmitted through the channels of the information system, whatever they are – plus the adjacent media, such as the translucent air, which transmits sound and light so well. You can finesse your ignorance of the gory mechanical details of how the information got from A to B, at least temporarily, and just concentrate on the implications of the fact that some information *did* get there – and some other information didn't. (1996, 359)

The puzzle here is familiar. Dennett wants a "multimedia" replicator, whose identity conditions do not rely on some underlying alphabet, which are not restricted to *any* particular or limited set of physical bases. Rather, these "informational entities" can take on indefinitely many forms while still being instances of the same meme.

Dennett's solution to this problem is unique, sophisticated, and tied deeply to his ongoing research project in the philosophy of mind. Unlike Hull, he does not have any faith that developing neuroscience will discover some physical similarity between brain-states that are instances of the same belief, much less does he expect that there is any underlying *syntactic* features or "structure passed on more or less intact" in all instantiations of the meme in different media (1996, 354f). He proposes, instead, that what individuates memes are not physical or structural properties, but *semantic* properties. Dennett uses the term "information" in this sense. What makes two cultural items instances of the same meme is that they carry the same information, by which he intends a semantic characterization of information.[12]

Ordinarily, at about this point one would expect an account of what semantic properties are, in materialist terms. This is, of course, not forthcoming. What Dennett gives us instead, in what will probably be his career signature move, is what he calls *The Intentional Stance* (1987). Rather than a theory of meaning (semantic properties), the intentional stance is a defense of the attribution of intentional properties to people, written words, and so on, in ordinary life as well in the human-social sciences. The intentional stance is one of a number of stances or ways of looking at things we may take on, like the physical stance or the *design* stance that we adopt when we engage in the analysis of biological adaptations. We assume the intentional stance any time we talk about what someone believes or wants, what a story is about, what a sign stands for, what a gene codes for. Each of these familiar uses depends on a sort of relationship that is still poorly understood, even rather mysterious. Our use of intentional language rests no more on a well-established theory of meaning than our ordinary (and even scientific) talk about physical objects rests on a well-grounded theory of what physical stuff really is. What grounds the legitimacy of various stances is their robustness and utility in use. Each stance has its own realm, and it is clear that our attribution of beliefs and desires is much more effective in predicting human behavior than any currently available physical theory of the causes of behavior. Arguably, it will ever be so.

This is not the place to pursue the analysis and critique of the intentional stance.[13] Suffice it to say that the identity conditions for Dennett's memes (what makes two cultural items instances of the *same* meme) are determined by our ordinary recognition of intentional or semantic properties, rather than by any physical or structural criteria whatsoever. Notice that this contrasts with the Dawkins-Williams "sliding" definition of the gene, which was an attempt to provide an abstract characterization of *which* structural or syntactic features constituted the gene's identity conditions. The intentional stance does allow Dennett to defer precise specifications of the meme's identity conditions. The problem with this strategy, however, is that in eschewing all structural and physical properties as criterial for instantiations of a particular meme, Dennett loses any right to make any prediction about how these things behave, or even whether they are objects in their own right or properties of something else (like the brains of human beings).

Why this is so takes a bit of explaining. The outline of the problem is simple enough, I suppose. Intentional properties may have explanatory and predictive value, but if they do, it is because they covary with physical or structural features of their "vehicles," which tend to have predictable causal consequences in particular sorts of systems. Intentional properties themselves are causally inert. Having no necessary physical properties, they can have no necessary causal consequences. In the parlance of causality, they are *epiphenomenal*. They don't actually *do* anything at all, strictly speaking. Indeed, they don't even *cause* us to recognize them – recognition must be due to the way their vehicles interact with our sensory and cognitive apparatuses. Now, I think we may have to grant Dennett's general point regarding the intentional stance – that talk about intentional properties really is robust enough to do the work of science,[14] and perhaps we may even have to accept his characterization of the intentional stance as a sort of symmetrical alternative to physical description, one with its own domain. Even if we give him these points, bereft of necessary physical correlates, there is no reason to think that memes are selfish or, indeed, have any predictable behavior at all. (This point is elaborated in the next chapter.)

What Exactly Is a Meme?

Dennett's frequent references to Dawkins seem to imply that he thinks that Dawkins's status as an expert in evolutionary theory legitimates his appeal to the meme, but this will not do. Dennett's rejection of *any* structural criteria for "memehood" cuts him adrift from whatever Dawkins may have established in terms of the role of smallest stable entities in sifting processes. Dennett has other resources, however. As far as justification goes, Dennett's meme concept rests on two foundations. First is the intentional stance, which allows the individuation of memes according to semantic properties. Second is the application of a standard sort of definition of evolution to the basic insight that imitation is a kind of replication. The definition is given as follows.

> The outlines of the theory of evolution are clear: evolution occurs whenever the following conditions exist:
>
> 1. Variation: a continuing abundance of different elements.
> 2. Heredity or replication: the elements have the capacity to create copies or replicas of themselves.
> 3. Differential "fitness": the number of copies of an element that are created in a given time varies, depending on interactions between the features of that element (whatever it is that makes it different from other elements) and features of the environment in which it persists. (Dennett 1991, 200)

The definition is quite similar to those given by Lewontin (1970) and Campbell (1974). Let's see how memes do, according to the definitions. Do ideas, for instance, have what it takes to get evolution going? (1) There seems no shortage of variation, human creativity and misunderstanding being what they are. (2) If imitation is a process of replication, then there is certainly replicating going on, although one might justifiably worry whether and to what extent ideas "make copies of themselves." If one is in the mood, however, one can "slippery-slope" such worries indefinitely by pointing out that nothing replicates independently of its environment; indeed, the way in which replication depends on environments is rather the point.[15] (3) Finally, as long as the features of ideas have some effect on their imitation, we will have "fitness" either with or without the scare quotes. So, on the face of it, the bare fact that imitation is a replicative process, combined with the abundant variation and obvious discrimination on the part of human imitators, seems to allow cultural imitation to quite handily satisfy the basic requirements. What is wrong with this conclusion is, again, not that evolutionary theory applies to culture, but the idea that replicators with their selfish tendencies are *necessarily* present any time evolutionary theory is applicable. I do think that evolutionary theory is general enough to apply to culture. I don't think the notion of replicator is a helpful way of characterizing this process.

As a graduate student, I had the privilege of taking a course in the philosophy of biology that was cotaught by the evolutionary biologist Francisco Ayala. As an aspiring philosopher, I was shocked and appalled to hear the eminent biologist announce in an introductory lecture on evolution that he was making up definitions of all the central terms – "evolution," "selection," "fitness," and so on – on the fly! Moreover, as a matter of principle, he did this in all lectures on the subject, including those to aspiring undergraduate biologists, although one rather suspects that certain formulations had popped up more often than others. Presumably, he was not speaking for the whole profession. Quite possibly the attitude expressed was not so much a matter of principle as a general unwillingness to take seriously the kind of definition-chopping in which philosophers so love to engage, much to the annoyance of our colleagues in the less abstract professions. (Philosophers get along much better with mathematicians on this head.) The lesson, at the very least, is that philosophers are apt to read rather more into the definitions of biologists than they warrant. Consequently, one need express no disrespect in suggesting that one cannot simply deduce the presence of evolution (whatever that means) from the satisfaction of the sorts of definitions that biologists find useful. They are, for the most part, not intended to function as parts of deductive systems. I don't intend this as a criticism of biologists. There is an art to knowing which details are important and which are not. Biologists have the luxury of *knowing* what they are talking about – quite literally. The context in which biological evolution occurs has certain regularities, some of which may be important and yet not make it into the criteria for evolution. So, for instance, when a biologist says that evolution requires sufficient heritable variation, it is assumed that this variation is random with respect to fitness, for that is the nature of genetic variation in life on earth. Relatively constant environments create selection pressures that are regular on a multigenerational time scale. This is also omitted from the definition, since that is how things generally are.

This is also the reason the Williams gene is not particularly problematic. As a type of object, it may be a bit peculiar in that one needs to know about all kinds of externalities, such as rates of selection, to know whether one has got one. So perhaps what Williams had in mind would be better characterized as a threshold than a sort of thing, but, in any event, it is only one step removed from a variety of clearly defined things – chromosomes, nucleotides, nucleotide segments, classes of nucleotide segments – that the ontological confusion is unlikely to constitute any real impediment to the progress of science. Anyone who builds their theory around the Williams gene can easily dispense with it in favor of bits of chromosome if the need arises. In the meantime, objects have a salience that thresholds do not, and such salience may help the dissemination of an important idea. Dawkins's original ambition was little more than to package for the general public Williams's point – that one simply cannot make assumptions about the accumulation of evolutionary benefit to "entities" which are too ephemeral to *accumulate* anything. The reification of the threshold of ephemerality into the replicator and the creation of the meme were part of the strategy of increasing salience; but where Dawkins (as a biologist) may, indeed, have known of what he spoke when he spoke of genes, the replicator as an abstraction (even in the "active germ-line replicator"; 1982) constitutes the generalization of an object which should not have been. When it came to the meme, Dawkins seemed quite willing to admit that he didn't know what he was talking about.

So, by the time Dennett gets the meme, it has lost whatever virtues the Williams gene may have had (because of the lack of a clear underlying ontology like the genetic alphabet). Dennett compounds this distancing by claiming that it is not causally efficacious physical (syntactic) features that individuate meme, but causally inert semantic or intentional properties. Now, as I said earlier, we may accept the general point of the "intentional stance," that ideas are admissible entities due to the regularity with which we can identify their presence and their (currently) irreplaceable utility in explaining human behavior. But what is it to say that these semantically defined classes are *replicators*, given that all past definition of the replicator had them individuated by physical or structural properties? Copying is not something anyone seems inclined to specify clearly, and Part II makes it clear that copying (whatever it may be) is not the prerequisite to selection and variation evolution that Dawkins makes it out to be. The sole substantial notion that flows from the replicator concept is that replicators are "selfish," and it is on this selfishness that Dennett seeks to ground the autonomy of the conscious subject.

The problem with Dennett's approach is not simply that without knowledge of the physical structures you can make no predictions about the evolution of the system. It is not simply the unwarranted assumption of features governing a dynamical system, but that the entities he defines are not even uniform with respect to the features that govern the behavior of that system. It is, after all, not the semantic properties of a meme but the physical properties of its "vehicle" that instantiate them that accounts for their effect on the behavior of the system. Dennett's memes need have no causal properties in common whatsoever and, as a consequence, *as memes* they can be assumed to have no uniform effects. Differing to this extent in physical particulars, one can say nothing about their fitness, for it is physical effects on which selection acts. Nor can one say anything whatsoever regarding the patterns of variation which are critical to the evolution of a system. Whatever sort of a scientifically respectable *thing* may be licensed by the intentional stance, it is not the sort of thing that can play the role Dawkins created the replicator to play in evolutionary theory. Replicators are supposed to explain functional efficacy, not rely on underlying functional uniformity in the way the intentional stance does.

Finally, this criticism of Dennett's meme should not be misconstrued as a mere rehash of the standard reductionist criticism of the supervenient categories of folk psychology. The problem here is not that intentional categories are not scientifically legitimate, but that one must be careful in their handling. In particular, one cannot, as Dennett does, assume that intentional similarity has the same sort of consequences as physical similarity. Dawkins's replicators, however short they may fall of the kind of entity that science can work with, at least have the theoretical virtue of physical uniformity. This justifies one in assuming that *if* one can specify them, *then* they will have some uniform effect. No such assumption is warranted for Dennett's memes. To be sure, one can describe their patterns of transmission and explain the reasons for transmission via the ordinary language strategies endorsed by the intentional stance, but intentional explanation offers only minimal clues to the underlying physical substrate on which the direction of evolution critically depends.

CONCLUSION

One might be a bit puzzled by the fact that none of the three architects of the most popular concept of cultural evolution had any real interest in the subject. Dawkins sought merely to make clear that his replicator was an abstract characterization picked out according to abstract principles. Hull made a considerable effort to get the metaphysics of biological evolution right, perhaps with an eye to straightening out the units of selection debate. In the end, however, he had little interest in the application of selection theory to cultural change, other than to predict that the effort of scientists would be directed to the promulgation of their ideas. What Dennett really needed to do was find a way to ground the autonomy of the subject and establish the special nature of the human mind within the framework offered by our current best science.

Then again, it is perhaps not so surprising that the most popular concepts are those coined to make rhetorical points, rather than those created to make a new kind of progressive scientific inquiry possible. Methodologically, ontologically, the meme is a mess. For the purposes of popular appeal, however, it could not have been better designed by a Madison Avenue advertising exec. I must nevertheless urge that the only relevance that the meme and its shortcomings *have* to the application of evolutionary theory to culture is as a *distraction,* or perhaps as an embarrassment. The good news is that there is good work being done on cultural evolution using formal tools adopted from population genetics and the modeling tools that computers make possible. This is the approach that I take, in Part II, to the construction of a unified framework for analyzing the interaction of biological and cultural evolution. Before turning to the *formal* characterization of that relationship, I would like to say a few words about how we should understand the *physical* relationship between genetic and cultural transmission and see if something clearer can be said about the notion of selfishness in evolution.

2

Ontologies of Evolution and Cultural Transmission

Every voyage of exploration and discovery requires specialized equipment and training. Disciplines that seem unusual may be essential to success. Familiar luxuries must be left behind. The course which seemed straight diverts around unforeseen obstacles, and naive expectations as to what would constitute success give way to what actually *can* be done. Fortunately for us, our lives do not depend on success of this expedition, but despite the fact that we are venturing forth in our armchairs, at our blackboards and computers, we can still get hopelessly lost if sufficiently unimpressed by the difficulties that lie ahead. It is possible to multiply confusions so that those who come after are faced with the double task of trying to figure out how to understand things *and* where exactly it was that *we* went wrong, and why.

Our little expedition is devoted to trying to understand the interdependency between biological and cultural evolution. We follow the map given by evolutionary epistemology, which suggests that there is some common, rigorous way of understanding evolutionary/Darwinian/ selection and variation processes such that the *same* process occurs on both levels. If the concepts are sufficiently general, then in principle we will be able to extend the analysis to systems with more than two interacting evolutionary processes. Starting in the next chapter, we see that the key to this level of generality lies in concepts of selection and variation which consist simply of a way of organizing causes of change, rather than requiring the discovery of what basic kinds of things evolve. Metaphysics gets left behind at the base camp in favor of the bare essentials required for technical climbing, along with heavy meals and books of poetry.

The mountaineering metaphor serves to express a certain attitude toward theoretical creativity. Our task is sufficiently complex that the concepts we employ need to be pared down to the essentials, optimized for functionality and generality. The last thing we need is concepts crafted for their connotations in theory wars unrelated to the task at hand. Dawkins and Williams wanted to kill group selection dead. They have not, by the way, been particularly successful at this (Sober and Wilson 1998), but they have instilled a bit of caution in the kinds of claims we make about the way group-level traits evolve, which may turn out to be a good thing. Hull was concerned to answer critics of science. Dennett wanted a materialist theory which allowed the right sort of autonomy to the mind. In the grand scheme of things we are all on the same "side" – in favor of a scientific worldview we can live with. Given our sharing of somewhat cramped quarters, however, we will get in each other's way occasionally, and the resulting elbow jabs should be understood as the somewhat irritated jostling for space normal among allies.

That being said, Dawkins's replicator concept has the virtue of driving home the point that entities which are too ephemeral to respond systematically to environmental selection pressures cannot be "units" of selection. If one wants to argue that something bigger than genes are benefiting from selection in a systematic way, one has a lot of explaining (and probably math) to do. The notion of a replicator serves to define a certain threshold which is relevant to the calculation of selective benefit. As such, the term seems to have found a home in the ongoing units of selection debate, although increasingly the "vehicles" and "interactors" are where the action is. Other than that, however, the concept is about as useful as the notion of "gene" on which it is based – which is to say, not very. Recall that George Williams defined a gene as "any hereditary information for which there is a favorable or unfavorable selection bias equal to several or many times its rate of endogenous change." Open up any biology journal or textbook and you will find that this is *not* how the term "gene" is being used. How could one use such a concept? One would have to be constantly checking on selection pressures to find out what the genes were in one's system. Rather, a gene is enough DNA to do a certain job, like code for a protein or enzyme, or perhaps even determine whether one has brown or blue eyes. This, one *can* study. We can ask, as the Human Genome Project does, how many genes are there? What do they do?

The difficulty in coming up with a clear, univocal concept of the replicator calls into doubt whether "the replicator" should be the thing we think about when we think about evolution as a general sort of process. For whatever the details of their definition, replicators are not a basic *kind* of thing. To be a replicator is to qualify for inclusion in a category, where that inclusion requires no sort of lasting intrinsic or relational properties. Rather, external contingencies conspire to ensure that the "structure is passed on more or less intact," that "the rate of change due to selection is several to many times that due to endogenous factors," or what have you. Genes, even if they are defined *functionally* as units of chromosome which are sufficient to do a certain job, are not basic kinds of things, but acquire their individual identities from some larger process of which they are a part.[1]

This chapter addresses two issues. First, it attempts to establish that we do, in fact, already have a common and respectable framework within which we can understand both biological evolution and cultural transmission without recourse to vague concepts of information or relationally defined basic entities. Second, it assesses the utility of the meme concept considered even as a way of looking at things. This question, I suggest, turns critically on the notion of "selfishness" and raises certain important issues regarding how people think about evolution in general. Let us begin, then, by stepping back and asking if there is any basic entity that we must say uncontroversially *exists* (rather than being an artifact of a perspective), that is always there when we look at life. Is there any unit that just *is* the basic unit of life, something that one can identify, count, without reference to the environment in which it occurs? The answer is yes: the cell.

AN ONTOLOGY OF THE CELL

It sometimes seems as though evolutionary theorists forget that there is more to life than genes, traits, and environments. Genes conspire to generate traits in environments, which in turn select between genes depending on how well the traits they generate do. What is left out of this schematic picture is the developmental process which centers around cellular division and tissue differentiation whose integration will be critical to the future development of a unified evolutionary biology. It is not my purpose here to attempt to engineer the overdue reconciliation between estranged cousins in the biological sciences (Raff 1996), nor is it to argue that the view which emphasizes the role of genes should be abandoned or denigrated in any way. Instead, I would simply like to point out that there are theoretical resources available in very basic biology which offer themselves for our consideration when attempting to put together an ontological "big picture" of biological evolution. This serves two purposes. First, it emphasizes another replicative process which underlies biological evolution and which constitutes what is, to my mind, a much more solid sort of metaphysical lineage than the genetic replicator. Second, this basic process admits of certain enhancements (horizontal transmission) which both add power to the evolutionary process and are prone to very particular kinds of pathologies. In later chapters I show that the adaptive dynamics approach does not involve any ontological presuppositions. Here I point out that the continuity of life over time is considerably more substantial than a lineage in which some sort of "information" is passed along and show how this offers a framework within which we can seriously ask the question of whether and when cultural transmission is "viral." I shall attempt to be concise.

The fundamental unit of life is the *cell*. All living things are composed of cells. The smallest living thing is a single cell. You and I are colonies or families of genetically identical cells (barring the odd mutation). The metaphysical lineage of which we are all a part, the "tree of life" which extends back a billion years or more, is fundamentally a lineage of cells. The relationally characterized genetic "replicator" depends on the continuity of the cellular lineage.

In the extreme, it may be that nothing can be truly said to "self-replicate," since all reproductive processes have environmental dependencies without which selection would not occur. Self-replication is thus a matter of degree, as evidenced by Dawkins's definition of the "active" replicator. Within this framework, cells are particularly interesting because of the relative autonomy with which they can reproduce. One relevant consideration should be the relative complexity of the reproducing entity and the environmental features on which it depends. Such considerations both provide a measure of autonomy and serve to identify units which increase organized complexity. Genetic replication, for instance, requires very particular sorts of environments (those within the cell). Without a metric, one cannot compare the complexity of a genome with the complexity required for its replication, but comparison to the environmental requirements of some unicellular organisms is suggestive. The bacterium *Escherichia coli* is routinely bred in petri dishes in laboratories on a bed of agar. Cyanobacterica, the so-called "blue-green algae" which were responsible in large part for the original atmospheric oxygen supply on this planet, require only light, carbon dioxide, nitrogen, water, hospitable temperature and pressure, and some minerals to grow. In contrast, genetic replication requires an intracellular environment which seems to be of the same order of complexity as the genome itself. "Self-replication" may indeed be a matter of degree, but if we are interested in identifying units that account for the localized increase in organized complexity that characterizes life, the cell is the place to start.

The logic of cellular reproduction is not that of the "original" and the "copy," but that of *binary fission*. When a cell divides, one is left with two "regrown" half-copies. The copying process that underlies the tree of life, then, is not that of originals and copies, but of binary fission and regeneration. The *something* that is passed on at each step is not some abstract entity, but half of the concrete original. Of course, after a few stages, one does get individuals that are completely new, but the point remains. The fundamental reproductive pattern is not "copying" that takes two steps to get a genuine copy, but repeated binary fission and regrowth.

Within the cellular lineage is another lineage, the lineage constituted by the fissioning and regrowth of DNA. A double helix, DNA has a symmetrical structure – two halves that are not identical, but isomorphic such that each is able, in suitable environments, to regrow into a new whole. Fission and regrowth are not distinct steps but happen simultaneously as the double helix "unzips." Nonetheless, the logic is that of the cell's binary fission, rather than of the copy machine, the scribe, or the imitator.

Nothing I have just said is the least bit controversial, with the exception of the emphasis I have placed on the cellular lineage. The processes I have described every biologist knows, even takes for granted. What's the big deal?

For philosophical purposes, the emphasis on the cellular process actually makes a substantial difference in how we understand the replicative lineage that underlies biological evolution. If self-replication is just binary fission and regrowth in a less complex environment, then the "something" that is passed along at each stage is not abstract, but a concrete half of a concrete original. Furthermore, the original itself is a concrete particular, easily identifiable under ordinary laboratory conditions. My argument is not that there are no abstract entities of the sort that replicator theorists embrace, but that for the purposes of understanding biological evolution, *they aren't basic*. Perhaps genes defined in the Williams-Dawkins way are entities in their own right. Perhaps not. Perhaps the question is one of practical ontology – a question of whether it is productive for us to look at things that way. There is, of course, no question whatsoever as to whether cells deserve a place in our ontology, and whether the Williams-Dawkins gene survives the harsh environment of our conceptual scheme – the functionally defined gene that is a part of a cell, which plays such an essential role in production of organic molecules and regulating cellular growth and division – will remain. In this view, genes do not constitute the basic stuff of life, cells do; but genes, for the most part, determine the heritable *differences* between cells (including their differentiating abilities) and, as such, deserve their central place in the study of evolution.

It is also worth reiterating that one of the biggest outstanding problems with the modern synthesis is its failure to date to fully integrate developmental biology. Standard evolutionary theory works with a genotype-phenotype model, which has a general tendency to treat phenotypic variation as being as random as genotypic variation is supposed to be. The result has been excesses of adaptationism (Gould and Lewontin 1978) which seem to ignore the way in which basic processes of cellular development constrain available phenotypes. Accepting a cellular ontology, while not a panacea for all the ailments of the modern synthesis, would at least constitute a recognition of what developmental biology has learned – that the cell is not just a convenient and theoretically dispensable intermediary between gene and trait, but also the basic and limiting substance with which biological evolution has had to work. One can perhaps understand selection without it, but one cannot understand patterns of phenotypic variation, the source of the very stuff on which the selection operates.

HORIZONTAL TRANSMISSION

If the fundamental ontology consists of lineages of dividing cells, then it follows that the fundamental reproductive process is asexual. While this may seem counterintuitive at first, especially to those used to a picture of "gene shuffling," it nicely prepares us to approach the issue of viral pathologies.

In the primary reproductive process, genetic material is passed "vertically" down the lineage of dividing cells. This is accomplished via the binary fission and regrowth of DNA during replication. It is worth noting that most reproduction, even in sexual organisms, is asexual. Human somatic (body) cells, for instance, are all reproduced asexually, and it is only in the relatively infrequent event of production of gametes that recombination or "crossing-over" occurs. In purely vertical transmission (asexual reproduction), the "interest" of all components of the genome are united via the shared future within the lineage. Individual genes do not assort via recombination in a population level "gene pool" but are confined to their own lineage. The phenomenon of the selfish gene does not emerge until there is sex or some other cross-lineage exchange. One could, of course, insist that genes are fundamentally selfish and that sexual recombination merely provides an outlet for this fundamental tendency, but one could just as well argue that genes are fundamentally cooperative ("fundamentally" meaning "in the ancestral asexual state") and that being cut loose from their brethren forces them to look out for themselves.[2] In either case, pure vertical transmission is quite well behaved.

If you look at life in terms of the genetic lineage, sex (fertilization) involves only vertical transmission of genetic material. Seen from the cell's-eye point of view, however, sexual fertilization requires "horizontal" transmission of the male's genetic contribution. Instead of genetic material being passed vertically down the lineage of fissioning cells as the female's contribution is, the male's contribution is passed horizontally between two cellular lineages. Sexual fertilization is not really a matter of two branches of the cellular tree "fusing," but of a special sort of cell (the sperm) being created to transport DNA to the ovum, sacrificing itself to its role as genetic messenger. (It is the much larger ovum which provides the cellular resources and machinery for cellular division.) Integrating standard sexual genetics into this picture is a matter of seeing how the particular mechanisms involved in the production of gametes (this is where recombination occurs) and fertilization allows shuffling of chromosome segments. This process is fairly well understood. What the cellular ontology highlights is that the shuffled genes that Dawkins wants to convince us are the paradigm for evolution depend on horizontal transmission, which is structurally distinct from the asexual vertical transmission structure of cellular fission and regrowth. Sex is an innovation which speeds up evolution by allowing the sharing of genetic success stories horizontally across nearby branches of the tree of life.

Not all horizontal transmission processes are as well behaved as sex, whose tendency toward meiotic drive is largely suppressed. As we learn more about the world of unicellular life forms, we are encountering a bewildering variety of mechanisms by which genetic material may make its way between individual microorganisms. *Streptococcus pneumoniea* can apparently take up bare segments of DNA from the surrounding environment and incorporate them into its genome through a process called 'transformation.' There is a process called "conjugation" by which pairs of *E. coli* bacteria may transfer genetic material via specialized appendages called 'sex pili.'

The most significant mechanism of horizontal transmission, for our purposes here, is through viruses (known as "phages" when they infect bacteria). A virus is generally little more than some genetic material with a protein coat, although in some bacteriophages more complex structures may develop. One of the more dramatic viruses is the T4 phage, which infects the common bacterium *E. coli*. Looking like Stephen King's version of a lunar lander, T4 settles on its victim, penetrates the cell wall, and like a syringe injects its genetic cargo into the bacterium. The unwitting cellular machinery proceeds to use all its resources, including its own genome, to replicate the genes and body parts of T4, assemble them, and burst, releasing maybe 100 new T4 viroids into the environment. This process, known as the "lytic cycle" is fatal to the host.

What sort of a thing is a virus, then? To begin with, they are not alive. Despite the impression one might get from T4, viruses are generally quite inert in any environment other than that within the host cell. They are not motile nor are they irritable; they do not metabolize. Viruses do not reproduce themselves; rather, they "trick" host cells into replicating them. Moreover, they are extremely specific in the kinds of hosts required; many can only parasitize a single species. There is, indeed, ample reason to question their autonomy. This point is driven home by the fact that the viral genes in general resemble their hosts' genes more than they do the genes *of other viruses!* Although the origin of viruses is unclear, some being quite ancient, the best guess currently is that viruses arise through the malfunction of standard-adaptive intracellular processes.

So, is a virus an entity in its own right, or is it merely a malfunction, a sort of runaway positive feedback loop in the intracellular machinery involved in the basic process of cellular fission and regrowth? I think the right answer has to be that it is not clear. Better yet, it is probably productive to look at viruses from both points of view. The specialized adaptations evident in T4 suggest that we can understand their accumulation via a history of variation and selection on T4, treating the host as part of the environment. The replicator approach emphasizes this. On the other hand, to fully understand a virus, one needs to understand the process from which it arose, the process which has gone wrong. The programmed responses of cells are what makes viral propagation possible. Intuitively, any organismic mechanism which responds to stimuli by producing more of the same may create a "chain reaction" with unanticipated consequences. Simply treating the host as environment does not help one understand the conditions under which viruses can emerge and evolve. Let me emphasize that my argument is not that memes are viral and viruses are not entities in their own right, but that the precipitous lumping together as "replicators" of genes and bacteria and viruses and ideas (memes) obscures differences that are essential to understanding each and their interaction with each other. Just as human health cannot be reduced to the balance of four fluids, however convenient that would be if true, there is more to understanding an evolutionary process than finding the replicator. The devil may be in the details, but so is the good stuff.

The importance of sex in speeding up evolution shows the power of horizontal transmission. Viruses make the dangers clear. If ideas sometimes behave like viruses, then perhaps there is some other (nongenetic) process of horizontal transmission that has broken down. There is good reason to think that viral ideas, if there be such, are even more intimately connected to ordinary adaptive processes of thought than viruses are to the cells they parasitize. What is critical for both memetics (which applies the replicator view to cultural transmission) and evolutionary epistemology is understanding the basic process which underlies cultural transmission. We will then be in a position to understand both the reliability of cultural knowledge processes and the factors which influence the extent to which ideas evolve for their own ends rather than ours.

Information and Meaning in Evolutionary Processes

CULTURAL TRANSMISSION AS A CELLULAR PROCESS

The basic ontology of life consists of the lineage of fissioning cells. The object in this section is to see how cultural transmission appears against the backdrop of the cellular ontology. We choose the cellular ontology as our backdrop for being minimal. Whatever sorts of things human beings are, they are unarguably at least families of cells. Cultural transmission as it occurs among human beings often consists of complex behaviors and responses on the part of the organism as a whole. Nonetheless, as far as basic scientific ontology goes, these are still cellular, or rather, *multicellular processes*.

One of the general features of life forms is that they are *irritable*. That is, they are equipped with mechanisms which cause them to respond to environmental stimuli in adaptive ways. This is true from the simplest bacterium up to the most complex multicellular organism. The motility of bacteria, for instance, is controlled by a variety of chemical sensors, resulting in both approach and avoidance behaviors. In Part II, I consider what it would mean for "information" to be transmitted through environmental interaction, but informational quantities turn out to be much more elusive (and dependent on your framework of analysis) than the causally efficacious physical and chemical interactions. Consequently, what is basic about such interactions is stimulus and response, rather than the "transmission of information," which is actually an assessment of system performance, rather than a causally explanatory *something* that is transmitted.

Cells respond to other sorts of stimuli than environmental attractants and repellents. In most multicellular organisms, genetically identical cells must differentiate into types, forming different tissues and organs. Experiments in which cells are transplanted from one location to another in developing embryos demonstrate that cells communicate chemically with each other coordinating cellular specialization. In tissue differentiation, chemical signals are transmitted horizontally between adjacent branches of the cellular tree. Again, perhaps one can give an informational characterization of this process, but that is a functional assessment of the role of such chemicals, rather than the process's nuts and bolts.

Development of a multicellular phenotype occurs over time, and this development is subject to environmental forces which result in variations in the structure of the phenotype, and thus in the fitness of the genes themselves.[3] Homeostatic mechanisms which regulate development by compensating for environmental fluctuations allow the structure of the resulting phenotype to be more precisely determined. This increased specificity in phenotypic structure reduces the random fluctuations in genetic fitness caused by unguided developmental variation and is thus evolutionarily advantageous. Homeostatic mechanisms also allow developed phenotypes to compensate for environmental fluctuations after the rapid development of the organism, facilitating the internal stability of the phenotype across a wider range of environmental conditions. Thermal regulation in warm-blooded creatures is just one of many such mechanisms. This phenotypic homeostasis provides the first clue to the biological basis of culture.[4]

Physiological traits are, of course, not the only ones that have fitness consequences and thus determine our momentary fitness. Behavioral traits do as well. For instance, reptiles accomplish thermal regulation via behavioral changes (moving toward rather than away from sunlight) rather than via physiological changes as we do. Reptiles are thus of variable *behavioral* types, depending on the state of their internal sensors. At a given time, they are either the type that approaches light and warmth or the type that flees it. Both the two behavioral dispositions and the mechanism that determines which of them is active are genetically determined in this case, a consequence of normal development.

What characterizes homeostatic mechanisms, compared with more elaborate controls on phenotypic variability, is their simplicity and their *compensatory* nature. Their function is to vary certain phenotypic characteristics in order to return internal states to the vicinity determined by the set point. As noted earlier, these phenotypic characteristics can be either physiological features or behavioral dispositions. Homeostatic mechanisms paradigmatically control a single process and operate so as to return the state of the system to stability in the short term; the selective advantage of such mechanisms is both enormous and obvious.

Although short-term stability is essential to reproductive success, and thus to fitness, longer-term survival and the details of reproductive behavior are no less so. Consequently, mechanisms which control phenotypic variability in response to environmental stimuli which have long-term fitness effects are also preferred by selection. Thus, if the shortening of the days stimulates an organism to change coats to a thickness better suited to cold and a color providing better concealment from predators against a snowy background, hoarding food for the coming winter, or hibernating to reduce consumption, mechanisms which cause such responses will be selected. Likewise, highly discriminatory responses such as those involved in mating strategies and "fight or flight" responses can have large effects on the proliferation of types characterized by such responses. All of this should be familiar and, indeed, rather elementary. The point is that behavioral and physiological, short-term and long-term mechanisms can all be understood under a common heading:

phenotypic variability as a programmed response to external stimulation. I propose to subsume cultural transmission under this heading as well.

When we acquire beliefs, we become different behavioral phenotypes, and these differences often have direct effects on genetic fitness. If I come to believe that drinking and driving is dangerous or that babies should be kept clean and warm, I become a type less likely to die in a car accident or to lose my child to neglect. In both cases, the behavioral dispositions associated with the beliefs have significant effects on the probability of my effectively passing on my genes, but the genes do not themselves determine the beliefs that I hold. What they do determine is that I am a creature capable of acquiring such beliefs, given sufficient conditioning or education. Presumably, the reason creatures such as I, capable of acquiring beliefs, are so common is that this capability confers selective advantage. Presumably, the costs involved in this capability are offset by the abilities of such creatures to acquire beliefs which result in the enhancement of their fitness.

The structure of acquired dispositions in humans is not as simple as these examples suggest. There are, for instance, higher-order dispositions. I may be disposed to believe whatever my older sister tells me, and if she tells me that drinking and driving is dangerous, I may become disposed not to drink and drive. My attitude toward my sister's pronouncements is a higher-order disposition because it affects my acquisition of beliefs and their attendant behavioral dispositions. Thus, the higher-order disposition's effect on my fitness is indirect, but no less profound for being so. Likewise, I may have acquired the disposition to take her pronouncements seriously due to an even higher-order disposition to listen to people who seem to be doing well in life. This in turn may have been acquired at some point, and so on. Nevertheless, we can still think of all of these dispositions, regardless of whether the associated response is characterized by acquiring some other disposition or by manifesting certain behaviors, as the result of programmed phenotypic variability, and evaluate them in terms of their contribution to fitness.

Human dispositions also tend to come in clusters and to operate in clusters. My belief that drinking and driving is dangerous (call it D) may be associated with not only a disposition on my part not to drink and drive, but dispositions to hide my friend's keys and call a cab when he is drunk, to attempt to convince others that cars and booze don't mix, or to put a M.A.D.D. bumper sticker on my car. On the other hand, if I'm feeling like doing something risky, perhaps drinking and driving will seem like just the thing to do. The disposition to do something dangerous and the belief D together make me disposed to drink and drive at that moment. Without either a disposition to take risks or avoid them, the belief D may not determine any changes in my driving at all. Thus we might want to say that *D* determines a disposition regarding whether to drive that is contingent on other dispositions, such as whether I am disposed to do dangerous things.

So human behavioral dispositions have several important features. First, there is a sort of hierarchical order determined by the relation between higher- and lower-order dispositions. Second, there is a tendency to cluster, both through having the states of other dispositions as conditions of activation and by groups of dispositions coming into and going out of existence together. The most peculiar thing about them, however, is that they are so often acquired through *imitation* (Sperber 1996; Tomasello 1999). Consider: in the general case of programmed phenotypic variability, a trait or disposition is acquired in response to some sort of stimulus. Generally, however, the programming or control mechanism, whether genetically determined or acquired in its own right, simply causes the acquisition of certain dispositions in response to certain stimuli. Imitation falls under the same description, but as a special case. *For when dispositions are acquired as a result of imitation, the type of disposition acquired is similar to that which stimulated the acquisition.*

The result is the phenomenon of cultural transmission via imitation. What we see is the proliferation of certain behavioral traits in a community, and it is natural to think of this as a kind of "transmission" and to wonder what kind of a thing ("information"?) it is that is being transmitted. The coordination of phenotypic variability that constitutes cultural transmission creates the illusion that there is some *thing* which is being passed and reproduced, much as the sequence of images on a movie screen creates the illusion that there are objects moving around in front of us, much as a row of falling dominos creates the impression of some thing that is passed along. Here, my point is to urge that the thing to do is to investigate the patterns of *proliferation* of acquired behavioral dispositions, what kinds of higher-order dispositions control this proliferation, and how the resulting dynamics succeeds or fails in optimizing gains in fitness, rather than inquiring into the nature of the elusive cultural replicator.

So here's the picture: human beings, like most organisms, are variable, multicellular phenotypes, where phenotypic variability is controlled by a variety of mechanisms. Selection ensures that these mechanisms have the net effect of increasing fitness, which is to say, environmental "fit." Human culture is part of such a control mechanism, although one of mind-boggling complexity. What characterizes culture is the mechanism of imitation, where behavioral dispositions are acquired in response to stimulation by the expression of similar dispositions. The importance of the concepts of homeostasis and phenotypic plasticity here is that they emphasize that as organisms we are capable of a wide variety of very specific state changes in response to very specific stimuli. This means that, although there are many instances of cultural transmission that involve copying proper (say, via part-by-part comparison), we clearly have the ability to undergo coordinated state changes in the absence of anything one might reasonably call copying. Moreover, clear instances of copying will be special cases of this more general imitative ability. So, we can account for the causal basis of cultural transmission and thus for cultural evolution without making any assumptions about the nature of cultural entities other than those revealed by the ongoing inquiry into the physical basis of human cognition.

The conservative characterization of cultural transmission, then, is as a kind of patterned acquisition of dispositions and is presumed to be adaptive, at least (1) as a general phenomena, and (2) in the historical context. *Cultural evolution is the dynamics of the distribution of acquired behavioral traits among human beings, with the emphasis on the role that imitation plays in governing this dynamics.* If the phenomenon as a whole is adaptive, it seems reasonable to suppose that the reason it is adaptive is that it governs disposition acquisition in such a way as to achieve a high degree of correlation between the distribution of acquired dispositions and their contributions to genetic fitness. Which is to say, if culture is, in general, adaptive, then the distribution of our beliefs should "conform" to the world in an epistemologically interesting way. On the other hand, the assumption of the adaptive nature of culture does not require that all instances of culture are adaptive. It leaves plenty of room for the kinds of pathological instances of culture that memetics tends to emphasize. Yet, even if one is convinced that there is (actually) no such thing as a meme, at least as a basic entity, there is still another defense of the meme concept that needs to be addressed.

MEMETICS: THE "MEME'S-EYE VIEW"

Dawkins and Dennett have been quite successful in popularizing the meme as a general concept covering the extension of evolutionary theory to culture. As of this writing, "meme" has been in the *Oxford English Dictionary* for more than ten years, there is a three-year-old electronic *Journal of Memetics,* and a small but enthusiastic community of "memeticists" dedicated to establishing a discipline around the "meme's-eye view." According to cofounder Francis Heylighen's 1998 editorial, "The Memetics Community Is Coming of Age," memeticists tend to be young, European, untenured, and *not* from disciplines such as sociology or cultural anthropology whose academic "turf" covers the patterns of cultural transmission that memetics seeks to explain. Conference reports speak of standoffs between "believers" and "nonbelievers." Among believers, often-heated debates about definitions and methodology share the spotlight with discussion of how best to transform memetics into a respectable discipline with funding and "good research." Notable as well are periodic claims that memetics will make the world a better place, by creating an antimeme meme which will teach us to defend ourselves against the ravages of the selfish memes, recalling earlier statements of Dawkins to this effect.

The growing body of memetics literature falls roughly into three groups. The first (Blackmore 1999; Dennett 1990; Lynch 1996) consists of attempts to use the suggestively "selfish" properties of memes to explain various patterns of the propagation of ideas and practices. The second group consists of essentially quantitative studies of cultural transmission patterns, which make little or no assumptions about the underlying tendencies of imitated entities to evolve for their own transmitability. These include work in ethology (Lynch et al., 1989; Lynch and Baker 1993, 1994) and resemble quantitative approaches to cultural anthropology (Sperber 1996) as well as pioneering work in cultural evolution (Boyd and Richerson 1985; Cavalli-Sforza and Feldman 1981). These quantitative studies seem to find "memetics" to be a timely label for an established and respected approach to the study of cultural evolution and transmission. The approach taken by quantitative studies is also of a piece with the modeling framework I develop in Part II of this book.

The third group consists of theoretical discussions of issues such as the definition of central terms and what the extent of the subject matter of memetics is.[5] The memetics community is generally cognizant of the kinds of difficulties I raised in the last chapter, but rejecting Dawkins's original implication that replicators are a prerequisite to evolutionary analysis does not seem to be an option. The emerging consensus seems to be that the emphasis on definitions is misplaced, which is probably correct. As I noted in the last chapter and will again in Part III, most important terms in science are not defined, and it is evident that the pursuit of precise definitions of operational terms can be paralyzing. Operationalizing concepts does seem to be what really matters, but this is not necessarily an easier task than defining them.

In addition, there are a number of popular (and somewhat premature, in my view) expositions of the "new science of the meme." This includes Dawkins (1993) use of "meme" theory to vilify religion and Richard Brodie's (1996) *Virus of the Mind: The New Science of the Meme,* a rather alarmist popular exposition of meme theory which attempts to draw a conclusion promoting critical thinking.

What distinguishes much of recent memetics from the efforts of Dawkins, Hull, and Dennett is the view that memes are not so much a basic "kind of thing" as a productive way of looking at cultural processes. Susan Blackmore's (1999) *The Meme Machine* is a measured attempt to get explanatory mileage out of this "meme's-eye view," applying the meme concept to subjects ranging from why we talk so much, why we have a big brain, to the nature of the self. She characterizes her approach like this:

> We can now start to look at the world in a new way. I shall call this the meme's eye view.... The point of this perspective is the same as the "gene's eye view" in biology. Memes are replicators which tend to increase in number whenever they have the chance. So the meme's eye view is the view that looks at the world in terms of opportunities for replication – what will help a meme to make more copies of itself and what will prevent it? (1999, 37)

Blackmore, like her colleagues, is cognizant of the various difficulties regarding the meme concept but defends her use of it on the basis of its utility in explaining phenomena of interest. Her application of the concept is not based on a metaphysical defense of the sort of things that memes are supposed to be; rather, she takes memes as providing a point of view which emphasizes important factors in the dynamics of cultural transmission which are hard to see on other views, like the one popular in behavioral psychology and economics that stipulates that people always act to satisfy their preferences or the now-unpopular sociobiological view that insisted everything people do is for the good of their genes.

How does one go about assessing a "point of view?" In the first place, points of view do have an important place in our understanding of the world. From a practical, functional point of view, many disputes which appear to be about how things really are or what kinds of things exist are really about how we should look at the world. Or, better yet, they are about how we should construct the framework within which to try to learn about that world collectively. Sometimes the answers are clear – some kinds seem natural. Cells, atoms, and organisms suggest themselves as such basic divisions of the world with which we interact that we are inclined to insist on them. Such basic stuff deserves a central place in our fundamental conception of how things are.

Yet, many times when we ask whether something *really* exists, the right answer is "perhaps." When the right answer is "perhaps," then perhaps we should really be asking about the fruitfulness of a point of view rather than nuts-and-bolts ontology. Are there really memes? Perhaps. Are ideas really just transient states of nervous systems and cultural transmission no more than coordinated phenotypic variability? They are at least that, and perhaps they are *just* that. Are viruses entities in their own right or are they particularly maladaptive traits of their "host" cells? Hard to say. Is sexual fertilization an instance of horizontal or vertical transmission? Depends on whether you are taking the cellular or genetic lineage as your point of reference. I fully believe that in each of these cases, we are not asking questions about matters of fact, but about how to look at things, *in order to be able to determine what the matters of fact are*. If facts are constructed, then the concepts we share determine which facts we can construct. In each of these cases, effectiveness in pursuing the task at hand may legitimately be a bigger factor in concept choice than our more realist allegiances to nature's true divisions.

The reader cannot help but be aware by now that I do not like the meme concept. It seems, in a word, "superstitious" to me – just the sort of concept that scientific progress will require us to abandon. People have a tendency to imagine entities where they should see patterns, personalities where they should see mechanisms. Memes, if we insist on their real existence, seem to be another instance of the excesses of human credulity. I see no way to prove that they don't exist, however. Proving anything requires something else that all parties hold in common against which one can get leverage. Given the nature of the rethinking we are engaged in, combined with the fact that the meme is still a moving target, leverage of the required sort doesn't seem to be available.

Fortunately, something can still be said. Ontologies are jealous, exclusive. Points of view are not. If memetics is, as Blackmore puts it, a matter of taking the "meme's-eye view," then this satisfies my major objection to memes. I do believe that there are cultural "contagions" that the memetic view effectively highlights, much as meiotic drive and viral evolution are highlighted by the gene's-eye view. On the other hand, I believe that knowledge is a matter of the adapted functioning of cognitive mechanisms, and while it is certainly possible to describe the adapted function of culture from the memetic perspective (appealing to some sort of evolved symbiosis between memes and genes), it is awkward to do so.

Tolerant pluralism is good as far as it goes, but it would also be helpful to be able to say something more specific about *when* the memetic and the more flat-footed coordinated-state-change views are each most likely to be illuminating. Ultimately, this will rest on the track records of the various approaches, but as it so often turns out, the clarification of a few basic concepts can give us some sense of how things might go.

THREE KINDS OF SELFISHNESS IN EVOLUTION

For the moment, let us set aside our worries and accept the replicator as a basic concept. Replicators are supposed to be selfish. (Recall that this was the ultimate bone to pick with Dennett's intentionally defined replicator.) What does this mean? There are actually several senses in which such things might be said to be selfish. I'll try to hold it to three. Genes, for instance, might be said to be selfish because they increase insofar as they *have self-beneficial properties* that promote their own increase. The emphasis here is on having the properties. Second, they might be said to be selfish because they increase at the expense of their neighbors, due to some recognizably *selfish trait*. The emphasis here is on the cost to others involved in achieving the benefit. Third, they might be said to be selfish because they have a tendency to *accumulate either self-beneficial or selfish traits*. The emphasis here is on accumulation. The three notions, then, are (1) *having* properties that promote increase, (2) having properties that promote increase in a selfish *manner,* and (3) having a tendency to *accumulate* properties either that merely promote increase, or do so in a selfish way. Dawkins's replicators were designed to satisfy the third sense, for that is what evolutionary theory – at least, as a general explanation for life on earth – is all about. On the other hand, much of the study of "micro"-evolutionary processes centers around the first and second senses.

The three notions are related in the following way. Anything that increases due to its environmental interactions must be selfish in the first sense. This is very broad. Anything that is selfish in the second will be so also in the first, selfishness is necessarily self-beneficial, at least in the short term. Anything selfish in the third sense must be selfish in the first as well. (Presumably it is possible to accumulate self-beneficial traits which do not impose costs on neighbors, thus sense 3 need not entail sense 2.)

A brief digression: Critics of the theory of evolution sometimes claim that the theory is vacuous. It is said that evolutionary theory claims that the fittest survive and then defines the fittest as the survivors. This is usually called the Tautology Problem. There are lots of answers. Sampling error (drift) may keep the fittest from surviving, even when fitness is defined as the tendency to survive and reproduce. This is not really a tautology (get out your logic book), but these answers are dodges. A better answer is that evolutionary theory may sometimes define fitness in terms of (the probability of) survival and reproduction; when it does so, however, what it claims is not that the fittest survive, but that current diversity and adaptedness of life on earth is the result of the historical accumulation of fortuitous accidents, where the explanation of the *accumulation* is that, well, the fittest survive. Or, to put it plainly, those that tend to increase tend to increase. The universe is populated by stable things. Call this tautological or circular or analytic if you like. Two points are critical. There is much more to evolutionary theory than the survival of the stable. The theory says that the survival of the stable combined with the minute instabilities that result in error in inheritance is sufficient for and is, in fact, responsible for the adaptedness and diversity of life. Second, the "tautology," properly called the principle of natural selection, is profound, pervasive, and if its formulations appear tautological, this is due to the inadequacies of expression rather than the vacuousness of the principle itself. Nevertheless, the frequent inadequacy of expression also demonstrates that we are prone to mistaking the explanatory import of the principle and the implications of its relevance. End of digression.

The problem with calling sense 1 "selfish" at all is that if things that survive due to properties that promote their survival are selfish, then everything that continues to exist is selfish. If we are not careful, we may fall into a position every bit as absurd as the "tautological" characterization accuses evolutionary theory of being. Of course, even on this account, not *everything* is selfish, only those things that survive. But if everything that survives is selfish, then attributing selfishness is no more than calling the thing a survivor. This is vacuous nonsense masquerading as explanation. I do think that the tautology has its place – for example, in the construction of a formal model of evolution, as shown in Chapter 3. But the bare fact of selection, even positive selection, implies little else either about the underlying nature or the predictable behavior of the selected thing. Consequently, all that Blackmore, Dennett, or anyone else is going to get out of the mere fact that ideas sometimes propagate due to properties that facilitate their propagation is that they can be expected to continue to be relatively stable as long as the environment in which they propagate does not change. The mere fact of successful propagation entails nothing whatsoever about the distribution of benefits involved in that propagation nor about the way in which variation arises and accumulates.

The second sense of selfishness is an economic notion. Replication in general is only sensitive to factors affecting replication (the "tautology" again), and sometimes it may impose costs that are quite high. Parasites often impose high costs on their hosts, and arguably certain cultural forms like the use of addictive drugs may propagate quite successfully while imposing debilitating costs on the addict. The distinction between selfishness in the first and second senses is subtle. Sober and Wilson describe it like this:

> We are accustomed to the idea that selfish individuals make groups less fit, so it makes sense that selfish genes make individuals less fit. Note that the word

selfish in the previous sentence was used to refer not to whatever evolves, but to units that benefit themselves at the expense of other units, within the next higher unit. (1998, 90)

So while anything that propagates effectively does so due to characteristics that facilitate that propagation, not everything that propagates does so at a net cost to the surrounding system. This is the difference between the first and second senses of selfishness. The question we need to ask ourselves here is, does it follow from the mere fact of replication that the replicators are selfish in this sense, or tend to be?

Clearly, the answer has to be no. Many genes are selfish in the first sense *by* being unselfish in the second; they propagate themselves by making positive contributions to the systems of which they are parts. Consequently, it does not follow from the mere fact of replication that the thing which replicates is selfish in the second sense.

The third sense is not often clearly distinguished, but the salience of the accumulation of self-beneficial traits often colors the notion of selfishness. Replicative lineages are not minds, or even homeostatic mechanisms. They do not compensate for the environment in complex ways, but in the simple way of competitive proliferation of variants. For the replicator to adapt, either to the same or a new environment, it needs to accumulate beneficial variation. More exactly, the lineage itself needs to accumulate such morphological variations. Consequently, the introduction of variation is essential to the third sense of selfishness, but not to the first two. But for a replicator (lineage) to be selfish in its accumulation of variation, certain factors need to be present which are not specified in the generalized definition of evolutionary processes.

First, the patterns of variation matter. As a matter of general principle, genetic variation is said to be "random" with respect to fitness consequences. The basis of this assumption is not conceptual, however. It is not built into the concept of variation that it be random in this respect. Lamarckian evolutionary theory, for instance, supposed that beneficial heritable variation can be elicited by environmental interaction. Lamarck's theory was not incoherent; it was simply false. It is matter of contingent empirical fact that genetics does not work that way, rather than following from the very concept of variation. Presumably, human genetic engineers can introduce variations in ways that are *not* random with respect to fitness, nor are they random with respect to the contributions of the modified replicators to the enclosing system. Moreover, whereas successful genetic variations tend to be small modifications on previous successes – due *precisely* to the random nature of that variation – if modifications are directed, then in principle there is no longer any reason to presume gradualist patterns of variation. Random variation seems to more effectively search the space of solutions in small steps. This need not be true for nonrandom variation.

Second, the environment is usually assumed to have a certain degree of consistency. If, however, the direction of selection changes faster than generation time, then the population cannot track it effectively, and there is no reason to think that new variants will be accumulated for any particular purpose.

So, for the common definition of evolution in three lines – variation, heredity, and selection – to be adequate as a specification of what it takes to get the familiar gradual accumulation of beneficial variations, we must make assumptions with respect to the pattern of that variation and the character of the environment. The requirements of variation and selection are necessary, to be sure, and, for beneficial variation to accumulate, something must ensure that the results of past selection are not lost when the ephemeral individuals perish. This is what heredity does. Omit any one of these three requirements, and the accumulation of adaptive variation will fail. The requirements may be necessary, but they are not collectively *sufficient* for accumulation. For accumulation, it is necessary that the environment does not dance around so much that the relatively sluggish process of differential reproductive success cannot track it. It is also necessary that the patterns of the introduction of variation are not such as to swamp the effects of selection.

What then of memes? What of the assumption that simply because ideas are imitated (which may fit some definitions of replication) that they have a tendency to accumulate beneficial variation? In the first place, it is only insofar as human reactions to ideas are uniform that the environment for ideas is constant. There are perhaps two sorts of reasons why this could be. Some ideas appeal directly to certain rather reflexive responses. People like social status. People like sex. People form addictions. This is all quite familiar. On the other hand, it may be that while an idea propagates by *appealing* to some response, in so doing it *satisfies* some set of standards according to which it is evaluated. Perhaps after all, some ideas propagate because they are true or useful. This would be the case when they propagate because they stand up to good tests of truth and usefulness. Furthermore, people can change their receptivity to an idea in much less than the time it takes to repeat it, especially if they are in critical mood. Now, it may be that there is a general principle of the sort that ideas behave more like viruses insofar as we are not discriminating, but that is not my point here. Rather, the point is that without knowledge of the particulars of the transmission process, we have no basis on which to assume that the environment for ideas is uniform enough to facilitate the systematic accumulation of beneficial variations. This is not to say that evolutionary models of the accumulation of variation are not explanatory for culture. On the contrary, there is good reason to think that the success of trial-and-error learning and the progress of science can be usefully explained in these terms, but mere presence of heredity, variation, and selection is not enough to imply the "selfish" accumulation of adaptive variants.

Similarly, there is no reason to think that the *patterns* of variation on ideas resemble those of genetics enough that we can assume that they allow for accumulation. Perhaps there are analogs to replication error, like the misunderstanding in the children's game where we sit in a circle and pass a whispered message around, but larger, calculated changes to an idea may be made in a single step by a single individual. Campbell (1974) argued that even individual creativity follows a Darwinian pattern, but that is a separate issue from whether ideas are replicators. We are, perhaps, more like genetic engineers and less like mistranscribed or shuffled genes in the modification of our ideas; which it is makes a difference.

I should like to be quite clear at this point that I have no quarrel with using the terms "evolution," "selection," and "variation" in an exceedingly broad way. Certainly, there are populations of ideas and to say that they evolve over time is just to say that the compositions of those population change. What does not follow from the applicability of the concepts of evolutionary theory is any expectation of *how* the population will evolve. It is not from the mere presence of variation, selection, and heredity that evolution tends toward higher fitness or greater adapted complexity. Critically, the way in which variation is introduced into the population and the way that selection pressures change over time determines where things go. In a nutshell, the mere applicability of the concepts of evolutionary theory is fully consistent with a system that chatters chaotically, darting one way and another, vaulting laterally via huge nonrandom variations, with no systematic behavior over time whatsoever. It is knowledge of the causal particulars which govern patterns of heredity, variation, and selection that warrants expectations regarding the behavior of an evolutionary process. It is precisely these causal details that are deemed irrelevant to memes construed in Dennett's fashion, which is the kind of approach one tends to take in the absence of an explicit theory. What all of this boils down to is that the theoretical motivation of the "selfish" replicator badly needs to be replaced by an analysis of what sorts of causal details actually bring about economically selfish propagation and the accumulation of variations.

LINEAGES AND POPULATIONS

Closely related to the distinction between kinds of "selfishness" is a certain ambiguity in the term "evolution" itself. Sometimes "evolution" means the transformation over many generations of the physical and behavioral characteristics of a species, and often the notion that this is accomplished via random variation and selection is included. Evolution either *designates* the actual historical process that has, in fact, resulted in the current diversity of life on earth or refers to a small class of highly similar processes. To investigate evolution is to investigate the details of that particular history. At other times, evolution means change over time in the composition of a population. Evolution in this sense may be a very short-term process. A population of peppered moths changes in its proportion of light and dark variants. The population is said to have evolved even though nothing new has been created. The term is used as it is in physics – evolution is change over time in a system according to a set of transformation rules. This is typically what is studied mathematically in population genetics and in evolutionary game theory. In short, one use of "evolution" emphasizes the importance of *morphological change* in a lineage, the other, the importance of the economics of *adaptive dynamics* of a population. Physiologists, systematists, and developmentalists (along with Ghiselin [1997] and Hull) tend to emphasize the first usage. Evolutionary theorists and geneticists tend to emphasize the second. Note that the second usage is the broader of the two, extending to cover the first as an instance.

These two emphases appear in our analysis of cultural evolution as well. When one applies evolutionary theory to the analysis of scientific progress, for instance, presumably what one has in mind is something like the long-term morphological change scenario. Successful theories accumulate minor modifications, which then affect the survival and propagation of the variants which carry them. Over the long haul, theories accumulate "adaptive" features via the distinctive Darwinian dialectic of variation and selection. On the other hand, epidemiological and memetic approaches to cultural transmission are generally much less concerned with the way changes to "memes" accumulate than with the patterns and causes of their propagation. This approach shares its emphasis with mathematical adaptive dynamics, rather than with the conception of evolution as long-term accumulation of adaptive variations.

In the analysis of adaptive dynamics, the accumulation of change is often left out entirely. In the standard practice, one assumes a certain starting mix in which all the competitors of interest are present at the outset. Causes of propagation are what matter, and these causes fall into two categories. First are the traits carried by the members of the population, which *do not change*. The composition of the population may change, but each individual is stuck with its properties. Second is the characteristics of the environment. Environments in effect pick out the preferred types from the assortment offered by the population by interacting with the traits they carry.

Thus, to study evolution may mean one of two rather different things. It may mean the study of the actual physical-metaphysical lineage that constitutes the history of life on earth. If that is what you are after, then sex and the attendant shuffling of genes is basic. So is the binary fission and repair process that characterizes what may be the truly essential process of biological "replication." On the other hand, to study evolution may mean to study the *general* principles which allow us to understand how it is that the *particular* structure of biological reproduction allows complex adaptations to evolve. This analysis proceeds much as it does in physics. One attempts to separate out various components of the process – selection, variation, heritability, environmental stability, vertical and horizontal transmission – and study these general forces in isolation and in conjunction, across ranges of parameter values. This part of the enterprise must be formal, heavily dependent on mathematical and computational tools, for ordinary language simply does not have the precision needed. If one studies concrete processes as in the first case, one can simply look at the process itself, its historical traces and its performance in the laboratory. Of course, one does need to quantify the results, but the numbers refer to observable quantities rather than to abstract principles. On the other hand, to study, say, selection as a general sort of process, one must create an abstract system in which only that aspect of evolution occurs. Such a system is generated by its definitions, and to be studied consistently by a group of researchers, the definitions must be clear. Mathematical definitions are the clearest for this purpose.

In the case of biological evolution, the two studies ultimately concern the same subject matter, and the generalized formulation of our understanding of what the biological lineage places demands on the formal theory. For instance, Hull's definition of selection as "a process in which the differential extinction and proliferation of interactions *causes* the differential perpetuation of replicators" and the standard population dynamics definition as "frequency change due to differential fitness" comes down to more or less the same thing, *when* we are talking about the basic biological process. When we turn to cultural evolution, the two approaches come apart. For the mathematical theory should apply to both, but our understanding of the particulars of biological reproduction may have little relevance to our understanding of cultural evolution, other than as another case subsumed under the same general theory. The general theory is the mathematical approach. The particular instances are the actual physical processes of biological reproduction and cultural transmission. Comparisons between the two will be illuminating, but only within the context of the general theory.

So when we turn to the application of evolutionary theory to culture, we have a choice to make. We can reason by analogy from the actual process of biological evolution and ask whether there are any processes in culture sufficiently like that process that we can use the analogy to make predictions and provide explanations. This is what the "meme's-eye view" does, and this is the approach taken by replicator theorists. I have tried to show why I do not expect this approach to be fruitful. The analogy is too weak to explain very much about culture. If one does nonetheless choose to take this approach, one must be clear which version of the meme one favors. There is not one meme-concept, but many, and they do not all have the same domain of applicability nor the same explanatory import. Moreover, they depend critically on "entities" whose operationalization can be expected to remain difficult. Semantic versions like Dennett's abandon any physical uniformity and thus any predictive or explanatory power for physical systems. More restricted versions may explain relatively little of cultural change; in any event, the mere fact of replication or the presence of selection, variation, and heredity by themselves entails nothing regarding either economic selfishness or the selfish accumulation of adaptive variations.

The other option, the one I explore in Part II, the one taken by evolutionary game theory and the quantitative approach to memetics, is to apply the mathematical understanding of evolution to cultural change. My ultimate reason for taking this approach is that it is selection as identified by *that* approach that facilitates transfer of information from environments into things like minds, both on the level of maintaining the reliability of the senses and in error detection in trial-and-error learning. I also am quite convinced that the mathematical approach leaves us on far firmer ground, both with respect to ontology to our ability to share and accumulate results in a way that is essential to a progressive inquiry.

If there are memes (if we choose to take them seriously), their environment consists in human beings, their critical nature, their affections, and their credulity. Considered as environments, humans are complex, which is why it is probably better to just not say that memes "self"-replicate. Sometimes, however, human responses are simple, predictable, and uniform. If we think of memetic propagation as a sort of "chain reaction," such a chain will be more robust when humans respond simply and uniformly. In agriculture, we know that limited diversity (monoculture) allows the wildfire spread of parasites. When human response patterns create a monoculture for the spread of an idea, then the critical factor in explaining the spread of the idea may be some feature of the idea that appeals to features of that uniform environment. It is in these kinds of cases in which the memetic approach is most likely to be useful. If individuals are not being discriminating in their responses, perhaps the memes will accumulate random changes, rather than those calculated to benefit the transmitter. On the other hand, if people's responses to ideas form a heterogeneous environment, then more is explained by those responses than by the particular features of the meme. In such cases, we need to be looking at what people are doing, rather than what their ideas are doing to them.

CONCLUSION

Science is an extension of the basic human perceptual and manipulative abilities. It is largely a matter of the investigation of *invisible things and forces,* invisible to the unaided eye, incomprehensible to the uniformed imagination, incomplete in the life of the individual knower. Science investigates the very small and the very large; the very fast and the very slow; the very precise effects of unseen forces. This requires the invention, distribution, and standardization of both instruments and *concepts*. I imagine us poking around in the dark with a stick. The stick encounters various sorts of obstructions and communicates a pattern to us. We begin to form concepts and later theories of the nature of the obstructions and then test these concepts and theories by further probing. Ideally, more than one person has a stick. Ideally, there is more than one sort of stick. This enriches our concepts and allows more complex predictions and more decisive adjudication between competing concepts or theories.

We find over time that certain concepts are useful, and others are not. But there are different kinds or grades of usefulness. Some concepts have purely personal utility – they make one feel better, they awaken certain feelings. We typically exclude these from any role in scientific explanation because they are not sufficiently objective. Observations regarding them vary too much from person to person to allow repetition and data sharing.

Even within the category of scientifically admissible concepts there are grades. The fiery element phlogiston was a useful intermediary which allowed quantitative comparisons between the disparate processes of combustion, calcification, and respiration. Pushed toward a central role in a more complete chemical theory, it had several shortcomings. For instance, it turned out to have negative weight. In France in the late 1700s, a better set of concepts was formed: oxygen, carbon, hydrogen, and the like. The new concepts were nicely developed to accommodate the fixed proportions with which substances combine, which gives us our notion of the atom. Atoms turned out not to be atomic in the sense of being indivisible, but they are basic in another way. Atoms are with us to stay, as are the basic categories into which they fall, the elements. They occupy a sort of stable explanatory nexus in a field of related concepts. Things very much smaller begin to get murky – perhaps there is a wave function which "collapses" when a measurement is made, perhaps we just haven't found the right way of looking at things yet. Get much bigger, looking at collections of very many atoms, and there are perhaps too many ways of looking at things.

I attempted to demonstrate earlier that if there is an "explanatory nexus" in the biological sciences, it is the cell. This does not, nor was it intended to, force anyone to abandon analytic frameworks that do not grant pride of place to the cell. The intention was to establish a sort of benchmark for claims of existence. A lot of conceptual questions come down to a matter of whether it is useful to look at things in some particular way. One finds alternatives and it seems that the only reasons for choosing between them have to do with the particular activities in which the combatants are engaged. Perhaps the existence of these "things" is a product of looking at the world in terms of them. If, in some sense, one wants to say that there really was phlogiston, surely it ceased to exist as such once chemists stopped believing in it. But not everything is like that. Some things are basic. Some things are always *there,* whether or not they are at the level of complexity of your current analysis. Now, perhaps this is because in locating the cell, we have succeeded in capturing nature's "proper" divisions, or perhaps it is because we have happened to light on some way of slicing the world which is so much in harmony with the kinds of things we need to do that it is shareable, robust, and indispensable – a permanent part of our ontology. Whatever the case, whatever your location on the realist-instrumentalist spectrum, reference to cells or reduction to cellular processes puts you on firm ground.

So in urging that we think about communication, imitation, and other kinds of cultural transmission as coordinated phenotypic variability in related multicellular organisms, what I mean to argue is that cultural transmission is *at least that*. Which is to say, any complete account must at the very least be compatible with the cellular-organismal state changes involved in cultural transmission. You can account for everything that memes are supposed to do in terms of things that human beings do. The converse does not hold.

The problem with this critique of replicators and memes is that it more or less sends us back to the drawing board in terms of how to understand Darwinian evolutionary processes in culture. This was, of course, the intent. Replicators and memes are really just heuristics, which may or may not help in the construction of a proper sort of a theory, but neither the focus on the cellular ontology nor understanding cultural transmission as I have suggested tells us how to understand cultural evolution. This is not the impediment it seems to be. Ultimately, the success of scientific understanding depends not so much on the concepts involved as by the measurements and models to which those concepts give rise. Fortunately for us, the formal, that is to say mathematical, theory of evolution is well enough advanced that we are not forced to look to the physical process underlying biological evolution and reason by analogy to what a similar process might look like in culture. We can, instead, generalize on the basis of the formal theory of evolution itself, examine those concepts, and think what it would mean to apply them to culture and its interaction with biological evolution.

II

Modeling Information Flow in Evolutionary Processes

3

Population Dynamics

SIMPLE SELECTION

Introductions to evolutionary theory commonly explain the concept of natural selection with examples like this: Walking along a rocky beach, you cast your eye down the shoreline and you notice that pebbles of different sizes, instead of being scattered at random about the beach by the crashing waves, are neatly arranged in bands according to size. This orderly arrangement, although it looks almost like the work of an intelligent mind (or at least a meticulous and industrious mind), is in fact the result of certain simple physical processes. Why each size finds itself stable in each range is, of course, a rather complicated affair, but why it is that similar pebbles are grouped together is not. Details aside, the basic principle is this: different sizes of pebbles have stable positions at different distances up the beach, and as each pebble lands in a range where it is stable, it tends to settle there. If it lands in a range where it is not stable, it is easily dislodged. Over time, this has an ordering effect, and although individual pebbles may be moved out of their particular "stability zone," the cumulative effect of the differential tendency of pebbles of different sizes to stay in different locations overwhelms the short-term chaos of crashing waves.

The universe, we are informed, is populated by locally stable arrangements and things. Things and arrangements that are locally unstable tend not to be much in evidence. The mechanism at work, natural selection, is so nearly trivial and so completely ubiquitous that it tempts one to accuse evolutionary theory of emptiness, or tautology. Of course, such accusations, as well as the various responses, are more than familiar by now. Simple selection processes – Dawkins (1986) calls them "sieves" – are, of course, only the beginning of the story for biological evolution. So, the story usually continues with the introduction of the notions of heredity, replication, and the "cumulative" change that results from the distinctive evolutionary dialectic of variation and differential reproductive success. Before long, sexual reproduction and the attendant shuffling of genes responsible for most current evolutionary novelty is introduced; somewhere along the line it turns out that replication is essential to natural selection (or to evolution, or both). If the author anticipates the reader's question, "what about the pebbles on the beach?" the answer is that the pebbles exhibit a sort of simple selection, which is by itself grossly inadequate to account for the wonderfully adapted (and coadapted) structures we see in modern organisms. Such processes and principles are important for purposes of illustration, but since we don't get very far in understanding biological evolution without the inclusion of heredity and replication, why waste terminology on them? "Natural selection" (and "evolution") is surreptitiously redefined in such a way as to include the necessary ingredients for apparently "progressive," cumulative change.

The consequence of this common explanatory development is, as I tried to show in Chapter 1, that when people turn to the question of cultural evolution – of the evolution of science, language, and knowledge in general – they tend to ask "what are the replicators?" If they have read further and learned of the "central dogma" of evolutionary theory – that only changes in the germ plasm and not in the somatic tissue are inherited – they tend to ask "what in cultural evolution corresponds to the genotype and what corresponds to the phenotype?" As I argued earlier, this tends not to get us very far. What we should do is go back to the pebbles on the beach.

The primary concern of this book is to try to understand knowledge from a biological point of view. More than that, we are concerned to try to find a way to study knowledge as a biological phenomenon, to "naturalize" epistemology, to turn it into a field of scientific inquiry. What does this have to do with pebbles on the beach? It turns out that the mechanism of information transfer that is essential to understanding the biology of knowledge is already evident in the pebbles case, just as the basic mechanism of evolution is. Of course, before we are done there will be plenty of twists and turns in the story of knowledge, just as there are in the story of the evolution of life.

You cast your eye down the beach and are struck by the orderliness, the beauty even, of the arrangement. It might have been designed by an intelligent mind (or at least a meticulous and industrious mind). But there is something else, something not just to admire, but to learn. The arrangement or spatial distribution of pebbles tells you something about how things are. It tells you that there is something at work that imposes order on the spatial distribution of pebbles. It also tells you locally, on *this* beach, where the stability zones are for different kinds of pebbles. What the repeated selection of locations has done is left the imprint of myriad local physical processes on the local distribution of pebbles. There is, in short, *information* about the local environment coded into the distribution of pebbles. If we knew more, we could, for instance, say something about the average height of the surf or the recent occurrence of storms.

This is not to imply that it is only selection processes which leave information in their wake. Physical processes in general leave traces, which may (if we know enough) allow us to infer back to their causes. Selection, as the net effect of a number of causal processes, does this as a matter of course. In the case of the pebbles on the beach, the kind of information it leaves in its wake is perhaps as trivial as the simple, single-step selection processes we are considering. Change the example: I watch the frequency of a dark variant of peppered moth rise while the frequency of the light variant drops. Given that this is the result of selection, what does this tell us? That this is the kind of environment where dark moths do better. Again, this seems pretty trivial. I acquire a new dog and discover after a year that the potted plants in my garden are all now in plastic pots, whereas originally there was an equal mix of clay and plastic. What does this tell me? That this is now the kind of garden where plastic pots do better. Plastic pots are more stable in the new environment. Again, trivial.

Aside from the triviality of the information transferred, you will now (or should, anyway) be worried about another problem. Aren't I importing the knowledge mechanism into these scenarios? If so, doesn't this beg the whole question of the naturalization of our understanding of knowledge? The answer is yes, although it is difficult to avoid presupposing knowledge of the world when building a theory. But try the following: one finds that in various regions, different styles of clothing predominate. In very northern or very southern regions, clothing tends to be heavier, and in tropical and equatorial regions, clothing tends to be lighter. Presumably, Alaskans *could* wear shorts and T-shirts, and Southern Californians *could* wear goose-down parkas, but they don't. From this, I conclude that Southern California is the kind of place where shorts and T-shirts do better, and Alaska is the kind of place where goose-down parkas do better. This seems like the kind of thing that might be good to know, a kind of nontrivial (or at least potentially useful) information. As for the second problem, forget about me for the moment. The information that Alaska is the kind of place where goose-down parkas do well is encoded in the dispositions to acquire and wear clothes of the locals and would be there (and would be useful) even if no "knower" ever observed the fact. That, I submit, is no trivial thing. There is no smuggling in of the "knower," either.

We still need a bit more apparatus if we want to account for something approaching knowledge, since it isn't clear whether the reason that this is the kind of place that favors *this* kind of clothing is due to the clothing being better for people (compensating for harsh environments) or because local trendsetters and determinates of social status make it advantageous. What we need for knowledge (or at least what I am planning to supply in what follows) is that the locals are a basically sensible people capable of "learning" via trial and error what kind of clothing is the best and that they *select* styles of clothing not due to fad but due to superior protection from the elements. Trial-and-error learning is easily understood as a cultural-level selection process, and when what is doing the selection is something like the acute discomfort that comes from dressing in a way that threatens to elevate or lower body temperature to a dangerous degree, then the distribution of clothing styles encodes valuable information about the local environments and that it is held in dispositions to dress makes it both exploitable and independent of third-party knowledge mechanisms. Reliable, learnable, exploitable information about the world is a decent candidate for knowledge, at least for the naturalist.

On the other hand, clothing styles are notoriously subject to the dictates of fashion. What this means from our point of view is that clothing styles inhabit a complex ecology. The tendency of humans to decorate themselves and to judge others according to how *they* decorate themselves frequently competes with more utilitarian factors in determining the distribution of dispositions to dress. The discomfort that arises from thermal extremes is only one of a number of selection mechanisms in the cultural ecology of clothing. But the distribution of clothing reflects how the world is, to whatever extent, because the mechanisms that *create* discomfort are reliable and the fact that that discomfort is correlated with risk of illness or death makes the information valuable. The story behind the reliability of the discomfort mechanisms is, of course, not a story of cultural evolution but of genetic evolution. *Our* ancestors did better because they were disposed to dress so as to keep their body temperatures within the "safe" range, and this is why we are sensitive to temperatures in the way we are. We need to understand both the cultural evolution of clothing and the genetic evolution of the discomfort mechanisms to understand to what extent *knowledge* about the world is encoded in the clothing distribution. Moreover, the two evolutionary processes interact. The evolution of genetically based learning abilities goes on *while* those abilities are influencing the course of cultural evolution. How well culture does in arriving at and maintaining solutions to problems affects, in turn, how well people with the aforementioned learning abilities do. Genetically inherited learning abilities affect the cultural "fitness" of cultural items such as clothing styles, and the local efficacy of cultural items such as clothing styles affect the well-being and reproductive success of the learners.

This "fitness feedback loop," along with the information-transferring tendencies of selection are the central components of the model of biological knowledge systems that will be developed in the next few chapters. The picture of multilevel selection processes is not a new one. Psychologist Donald T. Campbell, following on the work of Karl Popper, laid out the basic vision in his 1974 "Evolutionary Epistemology." Campbell's idea was that selection and variation processes are ubiquitous, occurring on many "levels" in biological and especially human knowledge systems, and it is via the interaction between levels that our beliefs and acquired dispositions get reliably and usefully correlated with the world. The challenge, if one wants to make a science of biological knowledge systems, is to try and make this clear.

MODELING EVOLUTION

There are a number of kinds of models of evolutionary processes. One of the best known is the cellular automaton, of which the most familiar example is the "life" simulator. The cells in a cellular automaton are not, as one might think, the dividing cells of biological growth. Rather, the cells are squares on a grid, much like a chessboard. The system evolves as follows: each cell can be in one of a number of states at a particular time. In the simplest examples, the states are just "on" and "off." At each time and for each cell, states of the (eight) surrounding cells are recorded, and then the state of the central cell is either changed or left the same according to some rule that dictates the "dynamics" of the automaton. In the "life" simulator, for instance, the rule is, "If two adjacent cells are on, then the cell stays in its present state. If three, then the cell is turned on; otherwise it is turned off." The process is then repeated, and this iterative cycle goes on indefinitely. As you probably know, this particular algorithm results in some very interesting patterns, in particular, the selective stabilization of a variety of distinct "life forms," and in an ecology complete with "predators."

Another popular approach to modeling evolutionary processes is the genetic algorithm developed by Holland (1975; see Mitchell [1996] for a good introduction). Genetic algorithms, as the name implies, model the evolution of actual strings of, well, numbers, although the numbers can be used to do all kinds of things. They could, for instance, function to determine the phenotypes of a population of simulated animals. Or, they could be used to set the controls on a complicated machine. Or, again, they could be used to determine the values of variables in a set of differential equations. Typically, one starts out with a modest number of such strings (say, a hundred). Each string is then "plugged in" to the application of interest – inserted as variables in the system of equations, used to set the controls on the machine, or used to generate a population of simulated animals. The performance of each string is then evaluated according to some test (the "selection mechanism"). Strings that do badly are replaced with variants of the more successful strings. Variations can either be "mutations," in which small random changes are made to successful strings, or they can be "recombinations," in which segments of two successful strings are spliced together to generate a new string. Each member of the new set of strings is then "plugged into" the application, evaluated, and selected. Again, the process is repeated indefinitely.

What cellular automata have in common with genetic algorithms is that both attempt actually to *instantiate* evolutionary processes in the computer. They model the behavior of actual individuals (strings of numbers or cells in a matrix) rather than the average behavior of a large population of such individuals. Cellular automata are particularly good for modeling spatial effects because of the inherent relationships of proximity that exist between cells on a grid. Genetic algorithms are particularly good for exploring large "solution spaces" and have come into their own as tools for finding solutions to systems of differential equations that are intractable under standard techniques. For our purposes, however, an older, simpler, and more common kind of evolutionary model will be more appropriate.

POPULATION MODELS

Mainstream evolutionary biology has not embraced the replicator as a central concept but continues to follow in the population-genetics tradition deriving largely from the work of R. A. Fisher, Sewall Wright, and William D. Hamilton. Within this tradition, the central object of analysis is the *population*. It is the population, rather than the lineage, which evolves, and that evolution consists of shifts in the relative frequency of types within the population. This *frequentist* or population-oriented approach has proved fruitful in extensions of evolutionary theory beyond the dynamics of pure genetic transmission, notably in Cavalli-Sforza and Feldman (1981), Boyd and Richerson (1985), as well as the large literature on evolutionary game theory following (roughly) from Maynard, Smith, and Price (1973). It is this mainstream tradition that we will be following in the development of the formal system in this chapter, although some care must be taken to avoid the pitfalls of analogical thinking.

In particular, the paradigm formalisms of population genetics[1] are geared toward the dynamics of sexual reproduction, which we should expect will not be appropriate for the more general class of evolutionary processes, including cultural evolution and, of particular interest, science as an evolutionary process.

Population models are a species of vector models. Vectors, to begin with, are simply lists of numbers. Newtonian physics uses vectors to specify the position and momentum of bodies. The position vector has three components, which give the x, y, and z components of the body's location. The general form of the vector is written like this: $\langle x, y, z \rangle$. The three-place position vector determines a point in three-dimensional space when the three variables x, y, and z are given specific values, like this: $\langle 3.1, 6, 7.8 \rangle$. The momentum vector also has three components, which give the current momentum of body in each of the three spatial directions. Consequently, the full-state space of the system has six dimensions: three for the spatial location of the body and three for the momentum of the body. The full state of the system, as determined by the values of the two vectors, picks out a single point in that six-dimensional space. The motion of the body is simulated by modifying the position and momentum vectors according to the laws of motion. Vector models thus represent the state of the system by giving specific values to one or more vectors, and they model the *evolution* of systems with sets of equations which modify those values over time, moving the point through the state space of the system.

Population models use vectors in a similar way. A population is a collection of individuals, categorized according to type. The distribution vector $\vec{p} = \langle p_1, \ldots, p_n \rangle$ gives the state of the population in terms of the frequencies of each of the **n** types of things in the population. Just as with the position vector in physics, this determines a point in the system's state space, which has **n** dimensions – one for each type of thing. Natural selection is simulated by changing the values in the vector according to the relative fitness of each kind of thing in the population. The evolution of the population appears as a trajectory across the population's state space.

Normally, the values of the vector $\vec{p}$ are restricted because $\vec{p}$ tracks the relative frequencies of types in the population. Frequencies, by definition, range between zero and one. Moreover, they all must add to one. This means that $\vec{p}$ is restricted to a subspace of its full **n**-dimensional space, a region with a distinctive shape. For a population with only three types (**n** = 3), the subspace looks like Figure 3.1(A): an equilateral triangle with vertices at $\langle 0, 0, 1 \rangle$, $\langle 0, 1, 0 \rangle$, and $\langle 1, 0, 0 \rangle$. This triangular region is referred to as the "simplex." The simplex for **n** = 3 is a *two*-dimensional equilateral

Information and Meaning in Evolutionary Processes

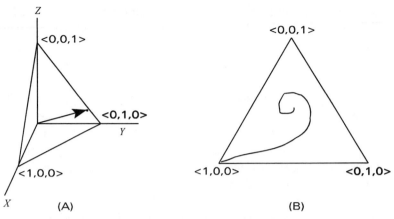

Figure 3.1. (A) Vector on simplex. (B) Trajectory on simplex.

triangle. The simplex for **n** = 4 is a tetrahedron, and so on. The simplex for **n** = 3 turns out to be a convenient way to represent the dynamic of population with three types (see Figure 3.1(B)). This is sometimes referred to as a "phase portrait." At the point in the center, all types are equally represented in the population. The vertices are points of "fixation" for each of the three types. On the edges the frequency of one of the types is zero.

As you might imagine, there are many sorts of population models. The ones we will be developing differ in important ways from the standard models of population genetics, as well as from the standard models of evolutionary game theory (the most "cultural" of the well-explored evolutionary models to date). Since we will be working toward models in which genetic and cultural evolution, along with their interactions, can be simulated in a single model, however, we need to build our modeling system so that it is general enough to cover both cultural and genetic evolution, learning, and inheritance. The trick is to start simple, and provide ways of adding features to the dynamical equations as they become relevant.

What all evolutionary population models have in common is *selection*. Selection consists of shifts in the relative frequency of types in a population due to differential fitness of the types. The basic mathematical representation of fitness and selection is simple. For each type i, there is, along with its frequency p_i, a fitness which is designated w_i. It is convenient to collect the fitnesses of the **n** types into a second vector $\vec{w} = \langle w_1, \ldots, w_n \rangle$. Think of the w_i's as "growth rates." Things that do not reproduce will have growth rates between zero and one. Things that do can try for higher growth rates. The way that fitnesses govern the dynamics of the population is simple as well. As a first approximation, the new frequency of some type i is the old frequency times the growth rate. This is written as

$$p'_i = p_i w_i. \tag{1}$$

At each generation, this multiplication is performed by substituting each of the 1 through **n** type indexes for "i" in equation (1), and this gives the new frequency distribution of types in the population. The problem with equation (1), however, is that there is no guarantee that the new sum of the p_is (written $\sum_i p_i$) will add up to one, which is to say, there is no guarantee that the new frequencies will actually be proper frequencies at all. The obvious solution is to divide each new frequency by the new *total*, a process referred to as "normalization." You could write this a number of ways, but the convention is this:

$$p'_i = p_i \frac{w_i}{\sum_j p_j w_j}. \tag{2}$$

The quantity in the denominator is just the new population size, but since the old p_is are assumed to be true frequencies, the quantity in the denominator is also the average or *mean fitness* of the population. Mean fitness is written $\overline{w}$, so that it is common to express equation (2) like this:

$$p'_i = p_i \frac{w_i}{\overline{w}}. \tag{3}$$

Equation (3) just says that the relative frequency of each type will increase if its fitness is higher than average and will decrease if it is lower than average. The greater the ratio of a type's fitness to the mean, the more that type's frequency will increase; likewise, if the type's fitness is smaller than the mean.

Suppose we have a population with ten types. Imagine the population to be whatever you like – varieties of hummingbird, clothing styles, or beliefs in various stories of creation. If we choose the initial frequencies of the ten types at random and do likewise for the fitnesses of each type, the population might evolve as in Figure 3.2.

Given that our distribution vector $\vec{p}$ inhabits a space of ten dimensions, and our fitness vector $\vec{w}$ a space of another ten dimensions, we need to be a bit clever to draw a picture of the state of the population. Figure 3.2 uses a pair of graphs to pick out a point in the twenty-dimension state space. The bar graph gives the values of $\vec{p}$, and the line graph gives the values of $\vec{w}$. This kind of representation is known as an "adaptive landscape," and the basic idea is due to Sewall Wright. Over the course of twenty-four iterations of our simple selection equation (3), we can see that type 4, which has the highest fitness, comes to predominate. Indeed, it predominates despite the

Information and Meaning in Evolutionary Processes

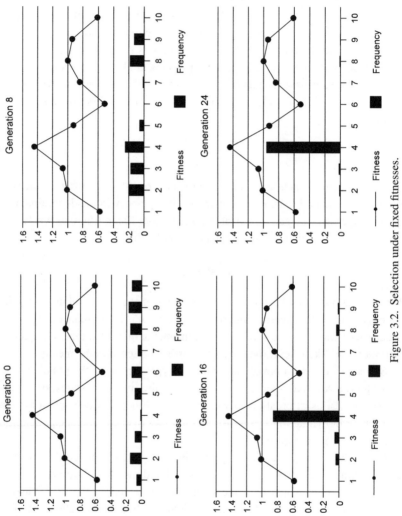

Figure 3.2. Selection under fixed fitnesses.

fact that it is initially not very well represented. In our example, the fitnesses range between .5 and 1.5. This means that the *absolute number* of individuals of a type can increase by 50 percent, or decrease by 50 percent in a single generation. The relative frequencies may not change at those rates, however, because it is the relationship to mean fitness that determines the growth in relative frequency, not the absolute growth rate. Indeed, the usual practice in population genetics is to adjust (or "normalize") the *fitnesses* so that the largest fitness is always equal to one. Here, however, partly because absolute growth rates seem more intuitive to me and also because we will want to use them in the multilevel models we will build in later chapters, we will continue to use absolute growth rates as fitnesses. The important thing to understand is that when the population frequencies are being normalized in each generation, as they are in equations (2) and (3), it is the relative size of the fitnesses that determine the dynamics, not the absolute value. For all the difference it makes, you could use fitnesses that ranged between 1 and 100, or between 1 million and 1 billion.

Figure 3.2 provides a useful heuristic for thinking about selection as an information-transfer process. It may help to think of the population as a simple learning device. What it learns is always basically the same: it learns which type of thing in the population has the highest growth rate in the local environment. Alternatively, what it learns is whether *this* is the kind of environment where type 1 has the highest fitness, or whether type 2 has the highest fitness, and so on. When the fitnesses are fixed as they are in our example, and no other sources of frequency change intrude, populations are reliable learners. Given enough time and a consistent environment, it is inevitable that the population will pick out the most fit type (4). The population does this picking out by being dominated by the most fit type, which, if you think about it, seems to make more sense than having it go to extinction. Of course, this is all metaphor, and fairly outrageous metaphor at that. The population isn't making use of the information or interpreting it. The thing to recognize is that the information is of a useful sort, and it is held in a useful way, *and* the gaining of this sort of information is inevitable in favorable situations. Things will get messier and more interesting soon enough.

Recall that we started by complaining that popularizers of evolutionary theory tend to focus on the specifics of biological-genetic evolution, thus encumbering attempts at a theory of cultural evolution with unreasonable prerequisites. Building a general theory up from the mathematics of evolution, on the other hand, leaves us much freer. Selection, from this point of view, does *not* require replication, so there is no need to look for replicators nor to despair over the prospects of a theory of cultural evolution when we don't find them. What evolution by natural selection requires is that there be a population whose type frequencies change over time due to fitness differences. Given that one can treat any collection of objects as a population, this reduces to the question of whether differential fitnesses are driving frequency shifts. But since a fitness is, by definition, a type's growth rate, however, it stands to reason that any group of objects that have fitnesses, where those fitnesses vary, will evolve by natural selection. So what kinds of things have fitnesses? There are a number of distinct notions of fitness, and these have changed considerably since Darwin. Many of these include the very notion of reproductive success (Dawkins 1982), but the notion of fitness at work in the mathematics of population modeling is quite clear: fitness is a numerical expression of a growth rate of types. Our choice of convention – that fitnesses are *absolute* type growth rates – makes things even clearer. It turns out that literally every type of thing in every environment has a fitness. Types whose members reproduce obviously have growth rates, but even types whose members do not reproduce have growth rates – it's just that the rates are always one or less. Types of nonreproducing things, strictly speaking, have measures of stability, or "decay rates." Mortality rates are part of reproductive rates, so it would seem rather arbitrary to insist that types which go out of existence at certain rates but do not reproduce have no growth rate, no w. There is certainly no mathematical reason for such an insistence, and since what interests us is how far we can legitimately go in applying the natural selection *equations,* there is no reason not to include decay rates as a special case of fitness. Consequently, we seem entirely justified in saying that every group of objects, unless every type of object in the group has the *same* fitness in the current environment, will evolve according to natural selection.

This is not to say that all frequency shift in every population is due to natural selection – that fitnesses subsume every cause of such shifts. Evolution, especially in the most interesting cases, proceeds by a kind of dialectic between selection and *variation*. The most commonly discussed sources of variation in biological evolution are genetic mutation and recombination, but importantly, *immigration* is also a nonselective non-fitness-related source of frequency shift in populations. Mutation and recombination are *endogenous* sources of frequency shift; immigration is an *exogenous* source of frequency shift. Yet, what is it that groups mortality (or stability) and reproduction as contributors to selection, and mutation and immigration as nonselective sources – contributors to variation? The answer to this question will allow us to distinguish *cultural* forces as contributors to selection or variation according to principle, rather than by appeal to analogy.

Population Dynamics

The answer to the question of what constitutes selection and what constitutes variation has two versions, one based on kinds of causal processes and one based on the mathematics. On the causal story, what makes something a contributor to fitness and thus a factor in selection (assuming fitnesses differ) is that the effect on the type's frequency depends on features of members of the type and the interactions of those features with the local environment. Suppose we have a population with types 1 through **n**. Type 1's frequency (p_i) decreases because of mortality – some of its members have died. This is an expression of fitness because it was members of type 1 whose features resulted in type 1's frequency shift. Type 2 increases due to reproduction. This is an expression of fitness because it is existing members of type 2 that result in type 2's increase. Type 3's frequency increases due to a massive influx of type 3 from outside the population. This is not a matter of fitness because the increase in type 3 has nothing to do with features of established members of type 3 interacting with the local environment. Type 3 did not *grow*, but was *added to*. A number of members of type 4 mutate into members of type 5. Consequently, type 5's frequency increases. Again, this is not a matter of type 5's fitness because members of type 5 didn't do anything to bring about the frequency change. Discussion of what to say about type 4's loss (i.e., is it selection or variation) will be deferred.

On the causal story, what makes a frequency change a matter of fitness-selection or a matter of variation depends on whether preexisting members of the type are involved in the change or whether they are "passive observers" of the frequency shift. This way of specifying the difference between selection and variation squares well with virtually all biological uses. It also turns out that, given that variation is simply characterized as sources of frequency shift that are not fitness components, then it follows that selection and variation cover all sources of frequency shift. This is an interesting consequence, in that it indicates that all sources of frequency shift can be accommodated within the conceptual framework provided by evolutionary modeling.

In the example, I said that after an influx from outside, type 3 did not grow, but was added to. The mathematical difference between selection and variation turns on the same distinction. In the causal story, in each case, the effect of fitness components on a type's frequency is proportional to the number of preexisting individuals of the type. A high reproductive rate won't make much difference in a type's frequency, at least initially, if there are only a few members of the type around. The way this gets expressed mathematically is that the measure of fitness w is *multiplied* by the old frequency. All selection processes in population models have this feature. Sources of variation, on the other hand, are characterized by something being *added* to a type's frequency.

The effect may, of course, be proportional to the frequencies of other types, but not to the type whose frequency is changing. So the easy answer, and as yet I have no reason to be dissatisfied with it, is that growth rates (fitness and selection) multiply frequencies, sources of variation add to them.

The general model of evolutionary processes that is beginning to emerge (and I expect that it may be viewed with alarm, or at least distaste, by some purists) is this: selection and variation processes are indeed ubiquitous, making the tools of evolutionary analysis applicable in a much wider variety of situations than has been imagined. Indeed, any arbitrarily chosen group of objects can be treated as a population. This does not mean that the population will evolve, for it can fail to evolve by not having the relative frequencies of types in the population shift. One would expect this sort of thing to be unusual, however. So we can say that any group of objects is a population, and most populations evolve. What about selection and variation? Every type of things has a fitness, a summation of "multiplicative" growth factors, in a given environment. Not every population that evolves does so by natural selection, however. Only when members of the population have different fitnesses does natural selection occur. Variation can cause the evolution of a population in the absence of selective forces (fitness differences), although one expects that the usual case is a bit of both.

What is potentially disturbing about this treatment of evolutionary theory is that it may seem to trivialize the concepts by making them apply to everything. I have some sympathy with this worry, at least insofar as it is motivated by the concern with trying to teach people how biological evolution *actually* works, a task which is frequently impeded by too free a use of metaphor. On the other hand, our aim here is to put together a set of conceptual and mathematical tools which will make tractable the analysis of the interaction between genetic evolution and cultural "evolution," and it just happens that the existing mathematical models are flexible enough to do this, with minor modifications.

Apologies aside, then, the reason that Campbell was right about the ubiquity of selection and variation processes is that the entrance requirements are fairly lax. In fact, rather than thinking about selection and variation processes as *kinds* of processes, it might be better to think about selection and variation models as *ways* of looking at or analyzing the evolution of populations. The appropriate analogy here is to statistical properties. The question in applying basic statistical concepts is not whether the population has a "mean" or a "median," but whether looking at the population in those terms is helpful. Likewise, the question of whether to use an evolutionary model to analyze a population is not whether the concepts *apply*. The concepts always apply. If the frequencies are shifting, then it is due to selection, or to variation, or more usually to some combination. It may be that no one has ever written down the kind of equations that determine changes in frequency in your population, but those equations, when they are written, will constitute selection, variation, or some combination of the two. This rather pragmatic attitude toward the application of evolutionary models is made all the more plausible if you consider that, in general, there is no clear answer to what the populations are, where one population ends and the next begins. Moreover, even if the population is given, there is still a rather pragmatic decision to be made as to how many types to divide the population into and, at some level, there may not be a single right answer. This is not to say that as an area of inquiry develops, we don't learn how it is most fruitful pick populations and types. This certainly seems to have happened in biology, with "island biogeography" providing a clear case of rather obvious population divisions.

The remainder of this chapter is devoted to explaining some of the standard ways in which the simple equations we have considered are made more interesting. I also add a nonstandard formula that will prove especially useful in modeling cultural evolution.

MUTATION

All evolution of populations consists of selection, variation, or both. We have already seen the most basic kind of selection model, and making selection more interesting consists mostly in making the fitnesses change in interesting ways. We haven't yet seen any models of variation or nonselective change, however. Recall that selection consists of multiplying a type's frequency by its fitness. Variation consists of adding something to the frequencies. The simplest model simulates *uniform random mutation*.

Mutation is a process by which an individual changes its type. This results in two things: the frequency of the type that the individual was a member of is decreased, and the frequency of its new type is increased. In uniform random mutation, every type mutates at the same rate, with an equal chance of mutating into every other type. Let **m** be the rate at which mutation occurs. A value of **m** = .001 means that, at each generation or cycle of the dynamics, one tenth of 1 percent of the individuals spontaneously change type. The frequency change, to begin with, is as follows:[2]

$$p'_i = p_i(1 - \mathbf{m}). \tag{4}$$

Of course, this only covers the *loss* to mutation of each type. How much does

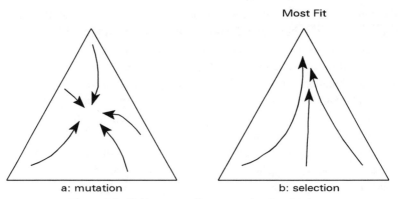

Figure 3.3. Uniform mutation versus simple selection.

each type gain? If **m** of each type is mutating, then it follows that **m** of the population is mutating. Because this is uniform random mutation, each type has an equal chance of mutating into any other (or back to itself). So, the gain to each type from mutation is $1/\mathbf{n} \times \mathbf{m}$, or $\mathbf{m/n}$, where **n** is the number of types in the population. Mutation *adds* **m/n** to each type, making the mutation equation

$$p'_i = p_i(1 - \mathbf{m}) + \frac{\mathbf{m}}{\mathbf{n}}. \tag{5}$$

Finally, we can combine our selection equation (3) with the uniform random mutation equation (5) to yield

$$p'_i = p_i \left(\frac{w_i}{\overline{w}} - \mathbf{m} \right) + \frac{\mathbf{m}}{\mathbf{n}}, \tag{6}$$

although we will find it more convenient in the rest of this section to consider mutation processes in isolation.

The effect of uniform mutation in isolation is, if anything, more predictable than the effect of selection under fixed fitnesses as in equation (3) and Figure 3.2. Figure 3.3 shows the difference for a population of three types.

Just as selection under fixed fitnesses always results in the most fit type taking over the population, uniform mutation alone always results in convergence on the state where every type is equally represented. This is because every type gains the same amount from uniform mutation, but those which constitute more of the population *lose* more. This tends to equal out the frequencies, and when selection and mutation are both occurring, this "flattening out" of the distribution interferes with the tendency of the population under selection to track the environment by becoming dominated by the most fit type.

Population Dynamics

This makes it sound as if variation is an impediment to selection, and it is true that when mutation rates are very high, they can lower the population's overall growth (mean fitness). On the other hand, sources of variation can do something that selection cannot – namely, increase the frequency of a type when it is zero. Fitness effects, remember, are always proportional to the frequency of a type, so that when that frequency is zero it remains zero, no matter how well members of the type would do if there were any around. Sources of variation, like mutation, being additive in nature (rather than multiplicative) *can* increase frequencies from zero. Look back at equation (5). If p_i equals zero, then p'_i will equal **m/n**, which at least gives it a chance to compete. There are a couple of other ways to put this. One is that selection never introduces the things that it acts on. Other processes, which necessarily fall under the category of variation, must do that. Another is that, under selection alone, extinction is permanent. If the frequency of a type ever falls to zero, there is no way that it can increase under selection alone.

Now, why is it that mutation, which is the paradigm of variation processes, involves multiplication as well as addition, translates into selection as well as variation? Mutation, as the term is used here, is a process by which an individual changes type. As noted earlier, this necessarily involves the reduction of the number of individuals of the old type and the increase of the number of individuals of the new type. The increase to the new type is a clear case of a nonselective process because the increase did not involve any established members of the increased type. It is the decrease that can cause confusion. It seems to me that clarity requires that we say that mutation, or type change, is a composite process which contributes (negatively) to the fitness of the old type and constitutes variation for the new type. Imagine that our population consists of chemical molecules and that type 1 is less stable in the current environment than type 2. Whether the destabilization of type 1 molecules constitutes "mortality" or mutation is simply a matter of what happens to the remains. If the remains of the type 1 molecule's destabilization are of type 2, then the process is mutation. If the remains are of a type that are not considered as part of the population, then the process is analogous to mortality. The cause of the type 1 molecule's destabilization is the same in either case – the interaction of the destabilized molecule with the local environment. The effect of the destabilization, the reduction of type 1's frequency, is also the same. So, there really seems to be no choice but to characterize mutation as a composite process, at least if mutation is defined, as we have defined it, as a process by which an individual changes type (and this is certainly what the standard mutation models simulate).

This may seem to do some violence to the biological usage of "mutation," but the tension seems to me to be instructive rather than debilitating. There are two reasons the selective aspect of genetic mutation is not ordinarily addressed in biological evolution. The first is that rates of genetic mutation are typically so low that they play a relatively minor role in the frequency shifts in genetic populations, serving mainly to introduce novelty in minute amounts. Moreover, genetic replication takes place in an environment that is well insulated from the environment to which the species as a whole adapts, so that whatever selective destabilization occurs in genes, it has little to do with the process of phenotypic adaptation that is the primary focus of the study of biological evolution. In short, the frequency reduction in standard cases is negligible.

Second, genetic mutations occur during the process of copying the gene, so that it is not clear exactly what to say about it. Is genetic mutation a process by which a gene changes type, or a process by which some new type is introduced *by* an individual of some other type, in an act of reproduction gone wrong? The reduction is negligible in either case; thus, there is no real reason to make a decision on this somewhat baffling ontological question. Nonetheless, one can see that in either; case, the loss must be, strictly speaking, a contributor to the fitness of the old type. For if it is a matter of the gene changing type, then the local instability of the gene that initiates the mutation is a fitness contributor. In the case of reproduction gone wrong, this constitutes a reduction in the reproductive rate of the old gene type – again, a fitness contributor. I am not suggesting that biologists should worry about this. All model-building requires simplification, and treating genetic mutation as a case of pure variation seems appropriate given that the negative effect on fitness is negligible. We should not expect, however, that in *all* cases mutation rates are so low – indeed, we will see cases in cultural models in which virtually all selective destabilization results not in mortality, but in type change.

In the uniform random mutation model (without selection) discussed earlier, the loss to mutation of each type (equation [4]) – although it was of necessity a fitness contributor – had no differential effect on the types' frequencies, because the mutation rate for every type was the same. In general, one might expect that mutation rates will differ between types and, moreover, that types do not mutate into one another with equal likelihood. Types which are more similar are more likely to mutate into one another than types which are quite different.

The usual way to broaden mutation models to accommodate these sorts of differences is to assume that each type mutates into each other at some fixed

Population Dynamics

Table 3.1. *Nonuniform Mutation Rates*

Mutation	Type 1	Type 2	Type 3
Type 1	.9992	.0002	.0006
Type 2	.001	.9987	.0003
Type 3	.01	.05	.94

rate. So to begin with we need a table or *matrix* of mutation rates. Table 3.1 specifies a mutation matrix for a population with three types. The row labeled "Type 1" gives the rate at which type 1 mutates into each type. Notice that type 1 "mutates into itself" at a rate of .9992. What this represents is not the tendency of type 1 individuals to destabilize and then return to being type 1, but just its tendency not to mutate. This was written as $(1 - \mathbf{m})$ in equations (4) and (5). So type 1's mutation rate is .0008, type 2's is .0013, and type 3's is a whopping 6%. There is, of course, an equation that goes with the matrix. The "self-mutation" rates in the matrix help make it simple. If we call the matrix M, then $M_{i,j}$ will be the contents of *j*th cell in the *i*th row. We can then write the following:

$$p'_i = \sum_j p_j M_{j,i}. \tag{7}$$

There are other ways to implement mutation matrixes in vector models, including those that combine matrix mutation with selection in a single equation, but this is the simplest. Equation (7) just says that the new frequency of each type (*i*) is the sum of the contributions of each of the various types (*j*) to *i*'s frequency, where those contributions consist of the old frequency of each type (*j*) times that rate at which it mutates into *i*.

FREQUENCY-DEPENDENT FITNESS

Matrices also prove useful in modeling changes in fitness. Although in the simple epistemological models developed in this book we will have little occasion to use fitness matrices, they will become essential later on as elements of competition and cooperation between cultural items move to center stage. The basic idea is that oftentimes how well a type does depends on how common various types are in the population. Fitnesses depend importantly on the environment (indeed, from the point of view of population models, fitnesses very nearly *constitute* the environment) and, for many populations,

Table 3.2. *Frequency-Dependent Fitnesses*

W	Cooperate	Defect
Cooperate	4	1
Defect	5	2

individuals in the population form an important part of the environment for other individuals. Frequency-dependent fitness models focus on the effects of these important internal interactions, assuming for the most part that external determinates of fitness, while they may differ between types, do not change over time.

Frequency-dependent fitness models have been used quite productively to investigate the evolution of cooperation. Suppose that a population of individuals inhabits an environment where they frequently have the opportunity to confer a benefit on some other member of the population, at a relatively small cost to themselves. Say the benefit is worth three units of fitness to the recipient, but only costs the donor one unit of fitness. It stands to reason that everyone will accept the benefit when offered, but only some will chose to confer the benefit. Following the standard convention, we will call those who both accept and confer the benefit "cooperators" and those that accept but don't confer "defectors." If we give every member of the population a background fitness of 2 (in this case, these are not absolute growth rates, but just convenient numbers), then this determines the payoff matrix in Table 3.2.

If we let W be the name of the fitness matrix (also referred to as a "payoff matrix"), then $W_{i,j}$ is the contents of the jth cell of the ith row of the matrix. The fitness of cooperators will depend on how many cooperators there are in the population – the more cooperators there are, the better cooperators do. We can calculate the fitness of cooperators by multiplying each payoff with the likelihood of running into each type of individual. If we assume that individuals in the population interact at random, then the likelihood of running into an individual of a given type is just the current frequency of the type. In this case, the expected payoff to cooperators is just $4 * p_{\text{cooperate}} + 1 * p_{\text{defect}}$. The mathematical expression for the fitness or expected payoff of an arbitrary type is $w_i = \sum_j p_j W_{i,j}$. Inserting this calculation of fitness into our selection equation gives us

$$p'_i = p_i \frac{\sum_j p_j W_{i,j}}{\overline{w}}. \tag{8}$$

In this case, $\overline{w}$ is the average of the new fitnesses as calculated via the matrix.

Population Dynamics

The game our population is playing is known as the "prisoner's dilemma," and the news is not at all good for cooperators, nor for the mean fitness of the population as a whole. The problem is that although cooperators do better when there are more cooperators around, so do defectors. In fact, no matter what the relative frequency of cooperators and defectors, defectors do better. Consequently, unless defection is actually extinct, it will inevitably drive cooperation to extinction. The result is that the population has a mean payoff of 2, rather than of 4 (which it would have if every member of the population had cooperated with every other).

As you may know, this result states a basic problem for the evolution of cooperation. Illustrating the problem mathematically like this, however, has made it possible for us to understand a great deal about how one might get around the problem. Indeed, the evolution of cooperation literature has been productive in generating different kinds of solutions to the problem of cooperation, which can then function as *empirical hypotheses* for real-life situations in which cooperative behavior has stabilized. My hope here is that by approaching the analysis of biological knowledge systems in a similar spirit, a progressive inquiry can be started in that area as well.

As I said, there are a number of interesting ways in which one can modify the model so that cooperation stabilizes. The most well known is the introduction of "sequential strategies," in particular, "Tit for Tat." The idea is this: one way in which our basic selection model is unrealistic is that we assume that individuals cannot choose whom to interact with nor *how* to interact. Sequential strategies are strategies for sequential interactions. In this model, we still assume that individuals interact at random, but that instead of just playing the game once with every individual they run into, they play the game a fixed number of times – say, five. Now, this repetition will not make any difference to the interactions between individuals that always cooperate and always defect. Enter Tit for Tat. Tit for Tat is just a little bit smarter than Cooperate and Defect. Tit for Tat can either cooperate or defect, depending on what the other player did in the previous round. Tit for Tat's strategy is to cooperate the first round and then on subsequent rounds do whatever the opponent did on the previous rounds. For a five-round sequence, this gives us the payoff matrix in Table 3.3.

You can work out for yourself the details of why this is the case, but it turns out that the presence of Tit for Tat can cause cooperative behavior to stabilize. The phase portrait for the Tit-for-Tat game is given in Figure 3.4.

It turns out that as long as the population starts out with more than 12.5 percent Tit for Tat, the population will evolve to some mix of Tit for Tat and Always Cooperate. Defection does well as long as it doesn't dominate the

Table 3.3. *Fitness Matrix with Tit for Tat*

W	Always Cooperate	Tit for Tat	Always Defect
Cooperate	20	20	5
Tit-for-Tat	20	20	9
Always Defect	25	13	10

population, but when Always Defect's numbers become large, the fact that Tit for Tat does better against Tit for Tat than All-Defection does against Tit for Tat becomes the deciding factor.

This is not the place to pursue the analysis of the evolution of cooperation. My main purpose in demonstrating how the Tit-for-Tat game works has been to convey a sense of how population models can facilitate the precise asking and answering of questions. The models are always simplified pictures of reality, but that very simplicity can be helpful in isolating mechanisms of interest. That simplicity also makes it easy to duplicate results. If you don't believe that Tit for Tat has this effect, see for yourself. Despite the relative

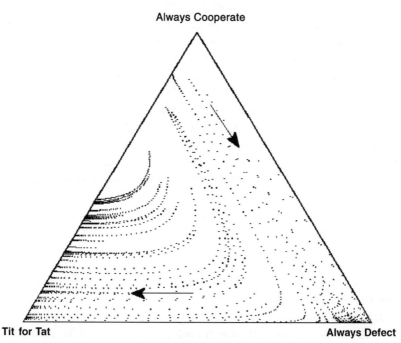

Figure 3.4. The Tit-for-Tat game with five iterated encounters.

informality of my presentation, I have given you everything you need to duplicate the model. Moreover, if certain simplifications involved bother you, the ability to duplicate *this* model makes it possible for you to extend *this* model. That makes it easy to see whether it was your modifications that had the significant effect, whereas if you built an alternative model from the ground up, this might not be so clear. My feeling is that this ability to duplicate results and incrementally add improvements to models is what lies behind the tangible progress we have made in understanding the evolution of cooperation. Obviously, I would like to see models of biological knowledge systems make the same kind of progress.

SELECTION-MUTATION IN FIXED-SIZE POPULATIONS

One of the problems we ran into in trying to apply the replicator idea to culture is that cultural entities, despite our tendency to think of them as tangible, are in fact more like patterns of behavior in populations of humans (or other animals) than kinds of things in their own right. We explored the question of how to think about their "transmission" in the last chapter, and it turns out that there is a nonstandard population model that will be convenient for modeling processes of phenotypic variability.

Suppose we are having a disagreement about where human beings came from, and you convince me via an ingenious argument involving the probability of beneficial mutations that the Darwinian theory cannot possibly be right and that the world and everything in it is, in fact, the result of a special act of creation by an all-powerful deity. The population of interest here is the population of beliefs about the creation of the world and the life forms in it, and what has just happened is that the frequency of beliefs in Darwinism has just gone down a little bit, and the frequency of beliefs in special creation has just gone up a little bit. If this were not just an isolated incident, but a general pattern, we would have to say that this seems to be the kind of culture where special creation does better than Darwinism and we could then wonder about why that might be. Obviously, the local culture is somehow discouraging the belief in Darwinism, but is the selection mechanism one that has to do with the ability of Darwinism to help us find our way about in the world, or is it due to factors more internal to our cultural life? Which of the two it is will have something to do with whether we say that I have "learned anything" from our discussion, but the immediate issue here is how to model this process.

Notice that unlike the familiar cases of genetic evolution, what happened was that at the very moment I changed my belief, one "token" of the belief in Darwinism disappeared and another token of the belief in special creation came into existence.[3] It seems natural to say that there are not two events here, but one. What happened is that my belief regarding the creation of life changed type – it went from being the type "belief in Darwinism" to being the type "belief in special creation." According to the terms we have been developing, what happened was that an individual belief changed type, which is to say, that it mutated. What is decidedly odd about this situation is that, unlike our mutation models discussed earlier in which there was selection on the "source" end to the mutation process and variation on the "destination" end, there seems to be selection on "both" ends here. This is the case because it was the inability of my belief in Darwinism to survive our conversation, combined with the attractiveness of your belief in special creation, that resulted in the mutation process. There is nothing "random" about this either. My belief mutated directly into a more (locally) fit belief just *because* it was locally more fit! Does the oddness of this general notion of mutation never cease?

Yes, I think this is the end of it. Mutation, as a process by which an individual changes type, is a composite process and needn't involve variation at all. There is always selection at the source end, and there can either be selection or variation at the destination end. That's it.

For our purposes, however, we won't need to model mutation with selection at both ends since we aren't going to model cultural transmission in this book. We have other fish to fry. What we do need to be able to do is model populations that are like our population of beliefs in the following respect: the population has a fixed size. The thing about beliefs about creation is that everyone has one, or at least that seems to be the most sensible way to model the population of beliefs. If anyone wants to claim that they don't have a belief, we will assign them to the type "none." If they are of divided opinion, we will give half of their full contribution to the belief-type frequency to one type, and half to the other.

The kind of fixed-size populations we are going to be looking at are ones in which every member of the biological population is always performing one of a number of behaviors, and when the behavior token "destabilizes," it changes at random to some other behavior (or back to itself). It turns out that this is a fairly good way of representing navigation in simple organisms such as bacteria and (to a lesser extent) bees.

In this kind of system (since there is no transmission), the fitness of a behavior is always one or less. Selection proceeds as before by multiplying the fitnesses w times the frequencies p, but because the population is of a fixed size, instead of normalizing after selection (dividing the frequencies by the new total), we *redistribute* the surplus generated by selection. Moreover,

Population Dynamics

since the "destinations" of the type change process are random, every type gets added to by the same amount, just their equal share of the total loss to selection (destabilization) in the population. The equation looks like this:

$$p'_i = p_i w_i + \frac{\left(1 - \sum_j p_j w_j\right)}{\mathbf{n}}. \tag{9}$$

Equation (9) just says that the new frequency of each type is the old frequency multiplied by the type's current local fitness (or "decay rate," if you prefer) plus the **n**th part of the population's total loss to destabilization. This dynamics has some interesting properties as we shall see later, the most important of which is the ability to track whatever is locally determining the fitnesses. It turns out that when the fitnesses are controlled by mechanisms that have been under selection at the genetic level, lots of useful information gets encoded in the distribution of behavior in an exploitable way. But there we're getting a bit ahead of ourselves.

SAMPLING ERROR OR DRIFT

The foregoing construal of populations assumes implicitly populations which are "effectively infinite," a phrase which, while common in population genetics, seems vague enough to be worrisome, and thus requires some clarification here. If one has a fair coin, then the chance of it coming up heads is 50 percent, as is the chance of it coming up tails. This does not entail, however, that for any series of flips the proportion of heads will be 50 percent. Rather, it is consistent with the coin being fair that a series of 100 flips will all come up tails. It just won't happen very often. How often is, of course, the subject matter of statistics. On the other hand, any infinite series of coin flips will (almost certainly) converge toward 50 percent heads. This is one way of stating the "law of large numbers."

Recall that our basic Fisher-style selection operator takes $\vec{p}$ to $\vec{p}'$ as a function of $\vec{w}$, such that

$$p'_i = p_i \frac{w_i}{\sum p_i w_i}. \tag{10}$$

The new frequency of types is the old frequency, times the fitness of the type, expressed as a percentage of the new population total. The fitness, recall, is just the expected value of increase for the type in the current environment, due to a certain subset of causes (those involving preexisting tokens of the type). So if the fitness of type *i* is 3, then, in all likelihood, the absolute numbers of *i*s will triple in each cycle. This does not mean, however, that each individual is guaranteed to have three offspring (to use a biological example). It is consistent with a fitness of 3 for *i*s to have a 90 percent chance of having no offspring and a 10 percent chance of having thirty offspring. Moreover, just as with finite series of coin flips, these likelihoods do not guarantee that in a group of ten *i*s, nine will have no offspring and one will have thirty. It is consistent with these probabilities that all ten have thirty offspring, or that all have none. What having a fitness of 3 does entail is that for larger and larger numbers of *i*s, the growth per cycle in the absolute number of *i*s will, with almost certainty, converge to 3. So, it is only in the case of *very* large, or "effectively infinite" populations, that there is any assurance that the population will behave as the fitnesses – which is to say, the environment, dictates.

What we are doing when we assume an effectively infinite population is simply ignoring sampling error for the purposes of the model. One might think that the only real excuse for making such an assumption is that it makes the models more manageable, and this it does. Sober (1984), however, offers a more theoretical justification for such assumptions. He characterizes evolutionary theory as a "theory of forces." This means that one undertakes the analysis of a system's dynamics by isolating different kinds of influences on the system's behavior and modeling their contribution to the dynamics in isolation. Newtonian mechanics is the paradigm here. We begin by modeling the effects of bodies in the absence of influences like gravity and friction and model the effect of those influences in isolation as well. We then proceed to the combined effects of those "forces" in more complex models. In evolutionary theory, the main "forces" that effect the evolution of populations are selection, mutation, recombination, and so forth. "Drift," or sampling error, is also one of the forces that acts on populations, to be understood first in isolation and then incorporated into more complex models as it becomes relevant.

We will largely continue to ignore sampling error in the remainder since its effect is mostly to introduce error into the informational dynamics that we are interested in here. The thing to remember is that the smaller the population, the larger the effects of sampling error, and thus the less predictable the dynamics.

There are two other explanatory features of sampling error that should be noted before moving on. The first is Sewall Wright's "shifting balance theory" of evolution, in which the very unpredictability of the evolution of semi-isolated subpopulations ("demes") serves as an important source of genetic variation for the more inclusive population. Hull (1988a), as discussed in Chapter 1, appeals to this theory as a justification for tightly knit subdisciplinary groups in science, although unsuccessfully (or so I argued). Likewise, sampling error may provide "noise" sufficient to break symmetries in the evolutionary process or to destabilize unstable equilibria (see Skyrms 1994). The point to keep in mind is that as our models become more complex, sampling error may allow us to account for otherwise mysterious phenomena.

The second point is that, according to the way we are constructing our models here, extinction almost always depends on sampling error. For except in the unusual case when a type's fitness becomes zero, our selection dynamics will never eliminate a represented type entirely. Extinction of a type is a matter of eliminating the last token of the type, and as long as we assume that there is no sampling error (i.e., that our population is effectively infinite), we can never be down to the last token of any represented type. Furthermore, the event which results in extinction must be a result of sampling error (i.e., failure of the type's frequency to follow the specified dynamics) except, again, in the unusual case where the type's fitness is zero. On the other hand, the dynamics of information gain that we will be exploring do not depend on extinction. Nevertheless, we should keep in mind that anytime we assume that extinction has taken place, sampling error has been at work.

CONCLUSION

This chapter began the task of building the tools required to model information transfer in multilevel selection processes. The central concept is that of a population – a typed collection of individuals. For such a collection, type frequencies may or may not change. If the frequencies change (including the introduction of new types), this constitutes evolution in the population. Evolution may be the result of selection, variation, or both. Selection is the result of differential fitnesses which are summations of each type's local growth and stability properties (e.g., mortality, reproduction). Variation constitutes the balance of causes of frequency shift, those not due to individuals of the type previously in the population (e.g., emigration, gains from random mutation). These schematic characterizations of evolutionary components are not intended, by themselves, to explain the accumulation of complex adaptations. *That* accumulation is the result of the constraints that specific causal processes place on nature of selection and variation in specific populations. Reproduction, for example, overcomes the natural loss of adaptive variation characteristic of collections of ephemeral individuals (e.g., pots in the garden) via the mechanisms of inheritance. The inaccuracy of those same mechanisms introduces new variants at low rates which are similar to previous successful individuals, facilitating the gradual exploration of the space of adaptive solutions. Thus, the concepts explored in this chapter provide a framework within which one can say what it is about inheritance mechanisms that are so ideal for accumulating adaptive variation. Those questions are tangential to the point of this book, however, which is concerned with finding a way of understanding evolution that does not place unnatural restrictions on our characterization of culture, but which allows us to capture the momentary adapting tendency of environmental interaction in its fullest generality.

The concepts of selection and variation that emerge are broad in their application and minimal in their requirements in a way that may make them seem to verge on triviality, but consider this: it is not having mass or location, but how much mass or what location an object has, that has explanatory power. Similarly, it is not having a fitness, but how much, and, more important, *why* a type has that fitness, that has explanatory power. Nor is it the simple fact that a population has sources of both selection and variation that explains anything, but why it has the particular sources it does and why those sources behave the way they do. The evolutionary concepts presented in this chapter do not and are not intended to explain anything. They are intended, rather, to provide a framework within which explanations can be made. Our explanatory use of them will come later. Moreover, they are for the most part commonplaces of theoretical biology.

I should emphasize one important novelty introduced in this chapter. Mutation, the process according to which an individual changes type, was characterized not as a pure source of variation but as a composite process in which selection (strictly speaking) operates on the source type resulting in the destabilization of the token and either random variation or selection increasing the destination type. (The example was changing belief based on an argument.) The latter possibility was forced on us by the need to accommodate populations of fixed size, which we can expect to be common in cultural-transmission scenarios. I argued that this should not be taken as a dispute with ordinary biological usage, in which the selection on source types is negligible and in which Weismannian inheritance ensures that mutational destinations are random with respect to fitness.

4

Information Theory

One thing that is missing from the formal framework for evolutionary analysis is a principled way of assessing the global-tracking efficiency of a knowledge system independently of the payoffs that drive its evolution. The purpose of this chapter is to propose a measure of *mutual information* borrowed from communications theory as an appropriate tool for measuring tracking efficiency, which is intended to complement and augment teleosemantics (see Part III of this book), standard evolutionary game theoretic formalisms, and Godfrey-Smith's (1991, 1996) application of signal-detection theory. It is important to note at the outset that the use of information suggested herein is different from the use that Dretske (1981) attempted to make of it in naturalizing meaning. In what follows, I begin with an introduction to information tailored to the needs of evolutionary epistemology. I then present reasons why this concept of information is appropriate, address worries regarding the metaphysical commitments entailed by its adoption, discuss briefly the attempt to use information theory to naturalize meaning by Dretske, and conduct preliminary investigations into the relationship between mutual information and payoffs in simple optimization processes. I close with a discussion of various concepts of information and their relation to the information of information theory.

INFORMATION BASICS

Although any thorough account of the history of information theory (also known as "communication theory") begins with the seminal contributions of Nyquist (1924) and Hartley (1928), things really got off the ground with Claude Shannon's (1948) "A Mathematical Theory of Communication."[1] Shannon was interested in problems concerning efficiency limits in telephone

Information and Meaning in Evolutionary Processes

and telegraph transmission. Using entropy functions to characterize the probability distributions of the states of sending and receiving devices, he was able to prove a number of theorems regarding the capacity of such "channels" to transmit information and the nature and availability of coding schemes to maximize information transmission. Since the boom in communications hardware in the early 1970s, Shannon's theorems have become invaluable tools for the communications engineer.

The needs of the epistemologist are clearly different from those of the engineer; the former is primarily interested in analysis, the latter, optimization. Consequently, I suspect that the real core of information theory – the coding theorems – will be of little use to us. What is of use, however, is the measure of information in terms of which these theorems are stated. I have something to say about the current enthusiasm for entropy measures and their metaphysical implications later, but the thing to do first is to lay out the conceptual and mathematical basics.

Consider two systems, **S** and **R**, which can occupy states $S_1 \ldots S_n$ and $R_1 \ldots R_m$. We want to know, in general, how well the states of **R** track the states of **S**, or *how much information* there is in **R** about **S**. This depends on how much there is to know about **S**, as well as on how accurately **R**'s behavior reflects **S**'s behavior. We begin by characterizing **S**. This characterization depends as much on how one describes **S**, which microscopic states we describe as distinct, as it does on the behavior of the system itself.

In general, the information generated by a system being in a particular state, called the "self-information," is

$$I(\mathbf{S}_i) = -\log_2 \Pr(\mathbf{S}_i).$$

Think of this as a measure of the uncertainty associated with the state $\mathbf{S}_i$. The behavior of self-information is characteristic of the logarithm. When $\Pr(\mathbf{S}_i) = 1$, then there is no information generated by **S** being in state i. If $\Pr(\mathbf{S}_i)$ is low, then the information generated can become arbitrarily large. So, information in this sense arises from the improbability or uncertainty of a state. The less common a state is, the more information is generated by the system being in that state.

The average self-information over all the states of **S** is what is called the *entropy* of **S**:

$$H(\mathbf{S}) = -\sum_i \Pr(\mathbf{S}_i) \log_2 \Pr(\mathbf{S}_i).$$

The entropy of **S** is a property of the *probability distribution* over states of **S**. Essentially, entropy is a measure of the unevenness of the probabilities

Information Theory

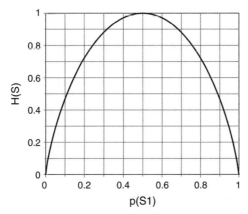

Figure 4.1. Entropy in a two-state system.

of states S_i. If the **n** possible states of **S** are equiprobable, then the entropy is maximized at $\log_2 \mathbf{n}$. Thus, entropy ranges from 0 to $\log_2 \mathbf{n}$ bits.[2] If, on the other hand, some one state occurs with probability 1, then the entropy and the information generated is zero. Generally, the flatter the probability distribution, the higher the entropy or average uncertainty. Figure 4.1 shows how the entropy of a simple system with two states varies, where $\Pr(\mathbf{S}_2) = 1 - \Pr(\mathbf{S}_1)$.

The "receiving" system **R** with its states $\mathbf{R}_1 \ldots \mathbf{R}_m$ has an entropy as well, and each of its states $\mathbf{R}_j$ generates $-\log_2 \Pr(\mathbf{R}_j)$ of information when it occurs, just as with **S**. To characterize the "transmission" of information between **S** and **R**, we need to know the joint probabilities of states $\mathbf{S}_i$ and $\mathbf{R}_j$, $\Pr(\mathbf{S}_i \,\&\, \mathbf{R}_j)$ as well. When **S** and **R** are statistically independent – intuitively, when none of the information in **S** and **R** is *shared* or "about" the other – the entropy of the joint systems is equal to the sum of the entropies of the individual systems: $H(\mathbf{S} \,\&\, \mathbf{R}) = H(\mathbf{S}) + H(\mathbf{R})$. When **S** and **R** are correlated, $H(\mathbf{S} \,\&\, \mathbf{R}) < H(\mathbf{S}) + H(\mathbf{R})$, the average uncertainty associated with the joint system is less than the sum of the average uncertainties associated with the individual systems. The amount of information shared $R(\mathbf{S};\mathbf{R})$ by the two systems, then, is just the difference between the two quantities:

$$R(\mathbf{S};\mathbf{R}) = H(\mathbf{S}) + H(\mathbf{R}) - H(\mathbf{S} \,\&\, \mathbf{R}).$$

This is known as the "rate of transmission" from **S** to **R**.

A more common approach to deriving the rate of transmission is to define the entropy (or average uncertainty) of **S** *conditional on* some state $\mathbf{R}_j$,

$$H(\mathbf{S} \mid \mathbf{R}_j) = -\sum_i \Pr(\mathbf{S}_i \mid \mathbf{R}_j) \log_2 \Pr(\mathbf{S}_i \mid \mathbf{R}_j),$$

over which we can average, giving us the average uncertainty about **S** in **R**

$$H(S \mid R_j) = - \sum_j \sum_i \Pr(S_i \mid R_j) \log_2 \Pr(S_i \mid R_j).$$

The average conditional entropy is known as the *equivocation* in the channel from **S** to **R**. Notice that the equivocation H(**S** | **R**) can function as a measure of tracking efficiency for the joint system.[3] If for each state of the sending device **S**, there is some state R_j which makes it certain, then **R** is a perfect indicator of states of **S**, given that there are as many states of **R** as there are of **S**. In such a case, H(**S** | **R**) = 0. (Perhaps a more apt characterization of equivocation is as a measure of tracking *inefficiency*.) H(**S** | **R**) maximizes at H(**S**) when **S** and **R** are uncorrelated. The rate of transmission then is

$$R(S; R) = H(S) - H(S \mid R),$$

which is just the entropy of the source minus the equivocation of the channel. The rate of transmission is always greater than or equal to zero, maximizing at $\log_2 n$ or $\log_2 m$, whichever is smaller. With a little algebra, the rate of transmission is equivalent to the "mutual information" I(**S**;**R**):

$$I(S;R) = H(S) + H(R) - H(S \& R)$$
$$= \sum_i \sum_j \Pr(S_i \& R_j) \log_2[\Pr(S_i \& R_j)/\Pr(S_i)\Pr(R_j)].$$

The latter is clearly a symmetrical relationship, which means that for mutual information there is no such thing as an inherent direction of transmission, apart from the causal particulars of the physical channel. Information in states of **R** about states of **S** is the same as information in **S** about **R**. Mathematically, mutual information is a symmetrical relationship resulting from the ratio of the actual frequency of joint occurrences to the frequency of joint occurrences if the states were statistically independent. (The product in the denominator is just the frequency of joint occurrences of statistically independent events.) The mutual information between two particular states can easily be extracted from the average form.[4] When S_i and R_j occur together, the self-information of the joint occurrence is $- \log_2 \Pr(S_i \& R_j)$, and the mutual information between the two systems at that moment is $\log_2[\Pr(S_i \& R_j)/\Pr(S_i)\Pr(R_j)]$ or the difference between the self-information of the actual events and the self-information of similar independent events $- \log_2 \Pr(S_i \& R_j) + \log_2 \Pr(S_i)\Pr(R_j)$. Note that via the definition of conditional probability, there are three equivalent ways to formulate current mutual information: $\log_2[\Pr(S_i \& R_j/\Pr(S_i)\Pr(R_j)] = \log_2[\Pr(S_i \mid R_j)/\Pr(S_i)] = \log_2[\Pr(R_j \mid S_i)/\Pr(R_j)].$[5]

Information Theory

WHY ENTROPY? WHAT METAPHYSICS?

There has been, from Shannon's first publication of the information-as-entropy formulation, a lively controversy concerning the formal identity of informational entropy and the entropy of thermodynamics, $-k \sum_i p_i \log p_i$. Some, like Cybernetics founder Norbert Wiener (1961) and Wheeler (1994), believe that the relationship reveals some profound truth about the nature of the universe. This trend sometimes brings in evolutionary theory as well, as in Brooks and Wiley (1988). Others, like Pierce (1980) and Wicken (1987, 1988), believe that the utility of equivalent mathematical formulas in thermodynamics and communication theory has no more significance than, say, the general utility of the Gaussian normal distribution or "bell curve" in a wide variety of disciplines. There is, as one might imagine, a variety of intermediate positions as well. (For a plunge into these deep waters, try Weber, Depew, and Smith [1988] or Zurek [1990].) Shannon himself, as early as 1956, warned that "Seldom do more than a few of nature's secrets give way at one time" (1993, 462), urging that a note of moderation be injected into the metaphysical excursions in information theory. Two questions arise: (1) Why should we use an entropy function to measure the information generated by a system? and (2) What are we metaphysically buying into through the use of information theory?

The answer to the first question is something one finds in any introductory information-theory text, as well as in Shannon (1948, 1949). If what one wants is a function f that measures the unpredictability, uncertainty, or *freedom of choice*[6] involved in the behavior of a system, it is argued that such a measure should have certain features. Most important is the additivity requirement. A complex choice or event should generate just as much information if it is analyzed as a single event or as a series of simpler independent events. Suppose we have two systems, **S** and **T**, that are statistically independent of one another, with **S** having two states (S_1 and S_2) and **T** having three (T_1, T_2, and T_3). Let us suppose that the two states of **S** are equiprobable, as are the three states of **T**. This makes the average uncertainty or entropy of **S**, $H(S) = \log_2 2 = 1$ bit, and the entropy of **T**, $H(T) = \log_2 3 = 1.585$ bits. If we consider the joint system **S** & **T**, then given that they are independent and the states of **S** and the states of **T** are equiprobable, then the six states of **S** & **T** are also equiprobable. Consequently, $H(S \& T) = \log_2 6 = 2.585 \text{ bits} = H(S) + H(T)$. This additive behavior of the entropy function also holds in the more general situation where the states of the independent systems are not equiprobable.[7] On the other hand, if **S** and **T** are not independent, then the sum of their entropies will be greater than their joint entropy, which is as it should be, since the

113

correlated behavior means that the comparison of the two systems reduces the uncertainty of the whole.

The standard argument goes like this: The additivity requirement already restricts f to a relatively small class of functions, one of which is the logarithm. The further requirement that f be continuous in the p_is and a decreasing function of the number of equiprobable alternatives narrows the possibilities to the entropy function

$$f(p_1, \ldots, p_n) = -k \sum_i p_i \log p_i,$$

where k is some constant. The choice of base 2 logs in information theory is arbitrary, although convenient. (Information units in base 10 logs are called "Hartleys," those in natural logs are called "nats.") There are a number of alternative formulations of the requirements with attendant proofs, but they all serve to show that if you want a few obvious properties, entropy is what you need.

The appeal of entropy as an information measure derives, then, not from the commonality with the formulas of statistical mechanics, but from the fact that it is arguably[8] unique in satisfying the intuitive requirements for a measure of uncertainty. It is simply the case that, if you want a continuous mathematical function for characterizing a probability distribution that increases as the number of equiprobable options increases, and adds in the way that a measure of the uncertainty of successive independent events ought to, you get a unique solution up to the constant multiplier k. This is not to say that there might not be something of deep metaphysical significance to be gleaned from the fact that both information theory and thermodynamics find such a measure uniquely useful, but the use of entropy measures surely bears with it no such presumption. Mutual information, as the natural extension of the entropy concept to the transmitted or shared reduction of uncertainty, is equally neutral.

The metaphysical neutrality of entropy, and consequently of information theory, becomes even more apparent when one considers that information theory does not *tell you* what states there are. Rather, without a predetermined set of states with probabilities to assess, one can't begin to evaluate either entropy or information. One has to bring a metaphysics or, at least, an ontology of states, to the analysis to get started. (More on this later.) A channel, in abstract terms, is simply a joint probability distribution. No more is presumed in the application of the mutual-information concept.

Because information theory is oriented to specific applications in communication technology, there is, of course, a component of the theory that deals with the structure of messages. If, for instance, one wants to *apply* the measure of information to problems regarding the optimization of telegraph transmission, then one needs to compare various ways in which messages can be composed of parts (letters), each of which has a number of possibilities. It turns out that, given a channel, there are optimal ways of setting up codes (i.e., the assignment of signals to letters), and upper limits to the amount of information that can be transmitted. One can also use the information measure to compare various ways of setting up the "information space." For instance, one might be able to increase throughput by reducing the number of different signals and increasing the rate at which they are sent.

This latter component of communications theory has also inspired metaphysical excursions. The idea of structured "information spaces" seems to have caught the imagination of Chalmers (1996), who suspects that information in "Shannon's sense" is the key to understanding mind-body dualism. What is troubling about Chalmers's treatment is that, following Bateson (1972), he characterizes "Shannon's" information as "any difference that makes a difference": "A 'bit' of information is definable as a difference that makes a difference. Such a difference, as it travels and undergoes successive transformation in a circuit, is an elementary idea" (Bateson 1972, 315). The problem with this construal of the notion of information is that it neglects the role of probability in determining information quantities. In fact, Chalmers seems to think that the information-theoretic notion of information is a matter of what possible states there are, and how they are related or structured (i.e., how the elements of states combine to determine the state), rather than of how probabilities are distributed among them.

The misunderstanding is a common one – for example, people often assume that binary devices such as switches always hold or generate one bit ($-\log_2 0.5$) of information. The truth of the matter is that binary devices generate a *maximum* of one bit of average information, but only when both states are equiprobable. If a switch is on half the time and off half, then it generates 1 bit. If, on the other hand, it is on 10 percent of the time and off 90 percent, then it only generates about .47 bit. That 10 percent of the time that it's on, it generates about $-\log_2 0.1 = 3.32$ bits (the same as the average amount it would generate if it had ten equiprobable states). If it's always off, it generates no information, just as if it's always on.

The maximum amount of average mutual information (e.g.) a light bulb can hold about a light switch is also 1 bit (given that we understand the bulb and the switch each to have two states, on and off). Perhaps it holds an average of 1 bit of information about whether the switch is on (if those events are perfectly correlated and equiprobable), but if your power goes out frequently, it will hold less than that on average, because switch-on and lights-on won't be perfectly correlated. Suppose, on the other hand, that your attic light switch was perfectly reliable but was only turned on three hours a year. Thus, the probability of light or switch on is $3/(24 \cdot 365) \approx .000343$. The entropy of either the switch or the lights will be

$$-.000343 \log_2 0.000343 - .999657 \log_2 0.999657 \approx 0.00445 \text{ bits.}$$

The average mutual information will have the same value, since the equivocation is zero by presumption of perfect correlation. When the switch and light are on that .000343 of the time, however, the bulb will contain 11.5 bits of information about the switch (and vice versa). When it is off .999657 of the time, they contain a mere .000495 bit.[9]

The frequent confusion about information capacities may arise from the fact that when informational concepts are presented, the first cases are usually those with equiprobable states. The motivation seems to be that in the absence of any particular knowledge of the nature of a source, equiprobability is presumed. Another reason that information theory may be thought to offer a structural notion (one centered on the arrangement of possible alternatives) of information is that information theorists are quick to point out that their sense of information should not be conflated with meaning. It is one thing to calculate the accuracy of sending and receiving signals; it is another thing entirely to say what those messages are *about,* or what it means to *understand* them. Consequently, one might think that since the notion is not semantic, it must be syntactic or structural. The dichotomy is false, however. What communications theory offers is a concept of information founded on a probabilistic measure of uncertainty. However, even respecting that information theory does not presume to quantify or explain meaning, there remains the possibility that the information-theoretic notion of information can be applied to semantic problems, at least if one thinks that covariance between the world and representational systems are relevant to those problems.

DRETSKE'S INDICATOR SEMANTICS

Fred Dretske thought that, despite the reservations of information theorists, there was a way of using information to develop a naturalistic semantics. In *Knowledge and the Flow of Information* (1981), he proposed that the conditional probabilities of the equivocation measure $H(S \mid R)$ were the key to understanding semantic truth conditions. The focus on equivocation to the exclusion of the source entropy makes sense in the context of semantics, since it is not the amount of information generated (i.e., the relative frequencies of states of the source) but the connection between source and receiver that is relevant to meaning. The basic idea was that, if the probability $\Pr(W \mid M)$ of a certain state of the world W given a certain mental state M is 1, then the mental state has that world-state as its content. In such a case, M contains the *information that* W – M is a reliable indicator of W. The stringent requirement of perfect indication allowed information to be preserved and passed along (satisfying the so-called xerox principle). If M contains the information that W, and M' contains the information that M, then M' contains the information that W also. If the requirements for M containing the information that x are less than $\Pr(x \mid M) = 1$, then this "transitivity" will not hold.

To the obvious worry that such perfect indicators are rarely if ever to be had in the messy world of biology, Dretske responded that the relevant probabilities were to be defined in terms of certain "channel conditions" (1981, 111ff). Roughly, the probabilities are defined for certain conditions under which the systems involved are operating normally. Thus, occasional mistakes are not necessarily enough to reduce the conditional probabilities below the necessary level. In later work, however (1986, 1988, 1995), the xerox principle was abandoned, and the requirements for indication weakened. Indication and "natural signs" or physical traces of events took center stage; information took on a supporting role.

The ultimate troubles for Dretske came from the need to accommodate misrepresentation (Dretske 1986). Suppose that we say that M represents W just in case $\Pr(W \mid M) = 1$. If this is the grounds of the representing relationship, the question is, how is it possible for M to *misrepresent* W? How is it possible to have a false representation if the represented state is guaranteed by the occurrence of the representation? Interestingly, it won't help to weaken the requirement and say something like that M represents W just in case $\Pr(W \mid M) > 0.99$. Of course, the weakened requirement does allow for failures, but a deeper problem remains: the problem of disjunctive referents. If the sign (a dot on a frog's retina is the overworked example) occurs when there is a fly there 99.1 percent of the time and when there is a berry hanging from a tree .9 percent of the time, then we seem to have satisfied the condition and can say that the content of the dot is something like "fly." The problem, as is well known, is that the disjunctive referent "fly-or-berry" also satisfies the criterion. What was needed is a principled way to eliminate the spurious disjunctive referent.

The problem of misrepresentation and disjunction has led Dretske and others to appeal to biological needs or biological functions to ground meaning. For Dretske (1986), representation becomes a matter of it being *the function* of an indicator to be a natural sign of something else. The dots on the frog's retina means (something like) "flying nutrition" just in case those dots are biologically supposed to be natural signs of flying nutrition. The proposed role for information theory here in explaining content has collapsed, since high degrees of reliability may play a relatively small role in determining biologically normative functions, as Millikan (1989) has been quick to point out. (Although, of course, it still might be the function of some system to hold information.) Godfrey-Smith (1996) has called into question the biological priority of maximizing $\Pr(S \mid R)$, as opposed to $\Pr(R \mid S)$.[10] It may be more important to indicate a predator whenever it is present – that is, maximize $\Pr(R \mid S)$, than to never get it wrong; that is, maximizing $\Pr(S \mid R)$. Consequently there is no reason to suppose, a priori, that maximization of $\Pr(S \mid R)$ will be the object of a representational system's function as derived from its selective history.

A further difficulty has been that the appeal to biological functioning alone is not enough to narrow down the class of potential representeds sufficiently. The function of a perceptual mechanism, like the frog's fly detector, is determined by what it has done in instances in the past where it has contributed to inclusive fitness. The frog's fly indicator has certainly responded to the presence of flies in such instances, but it has also responded to the patterns of light entering the cornea. Why is it the fly, rather than the light pattern that is the content of the dot on the retina? Dretske's suggestion was that multiple indication mechanisms might somehow "triangulate" on the right state of affairs, and that associative learning processes may solve the problem. An alternative diagnosis is that Dretske's focus on indication mechanisms is the source of the problem, a suggestion I explore presently.

My intent here is not to criticize Dretske's project, but rather to emphasize that, even in the early phases where conditional probabilities took center stage, he never exploited the resources of the full measure of information provided by information theory. What is distinctive about information theory is not simply conditional probabilities, which it shares with Bayesian approaches to epistemology, but the symmetrical joint entropy formulation that is exhibited by the mutual information formula.

USING FUNCTIONS TO DETERMINE INFORMATIONALLY RELEVANT STATES

Millikan's (1984, 1989) reaction to the difficulties of narrowing down potential referents has been that indicator theories of meaning make a mistake in trying to determine reference purely by looking at the (normative) conditions for *production* of representations. She proposes that we think of representational systems as composed of coadapted representation producers and representation consumers. Dretske's approach focuses purely on production and its evolutionary payoffs. Millikan suggests that it is consumption of representations – behavior on the basis of them and the reasons for the success from those behaviors – that determines meaning. So the dot on the frog's retina means (something like) "fly," rather than light pattern – not because it is the function of the retina to respond to flying insects, but because the dots on the retina have enabled the frog to get fed. It is, of course, a function of the retina to respond to the presence of flying insects (among other things), but the precise content of the dot is determined not by this function, but because it is the function of those dots to get the frog to eat flies.

Now, these subtleties are somewhat far afield from our discussion of information as a measure of epistemic success, since it should by now have become clear that meaning cannot be, without a lot of help from historical functioning, a matter of the kinds of conditional entropies that information theory trades in. My point in discussing the *behavioral* consequences of representation consumption is that a similar trick may help narrow our choices of ontologies of states, prior to the application of an information measure.

Recall that when we characterize our systems S and R, it is not given to us what the states $S_1 \ldots S_n$ and $R_1 \ldots R_m$ are, nor how many of them there are. This, while making it clear that the use of mutual information entails no ontological commitments, leaves us in the uncomfortable situation of having to characterize the states of the system arbitrarily. Millikan's suggestion, translated into this context, is that we can differentiate states of the environment S according to the behavioral options of the organism whose representational system R in which we are interested. Beavers, for instance, splash the water with their tails to signal danger to their conspecifics, which respond by swimming for safety. Because there is only one signal-behavioral response, we can say that R has two states, splashing and not splashing, and the vast number of states of the environment can be relevantly (from the point of view of the beaver colony's response system) divided into environments (states of S) in which it is best to splash or run and those in which it is not. Notice that we are using the differentiating power of selection – fitness difference – to differentiate world-states. From the point of view of evolutionary analysis, environmental differences that do not result in selection are invisible to the process of the adaptation of flexible response systems. Thus, it is fitness-differentiating world-states that are relevant to the characterization of the adapted response systems. So, although the appeal to the biological functions of signaling and representing systems does not help bridge the gap between information and meaning (the relationship of function to meaning being relatively independent of precise covariance), it may provide a principled way to determine the relevant states of the environment and representational system.

A TRACKING EFFICIENCY MEASURE FOR NATURALIZED EPISTEMOLOGY

The problem of misrepresentation makes compelling the point that *meaning* is not simply a matter of the statistical correlations in which information theory trades since having a determinate meaning is consistent with a high frequency of false representation. Representational systems emerge from the process of selection not for meeting arbitrary efficiency criteria, but for being good enough to beat the competition. This, however, does not mean that information theory is irrelevant to the naturalization of epistemology. On the contrary, reliability measures in the form of conditional probabilities have and will continue to play a straightforward role in the construction of models of the evolution and optimization of knowledge systems. I want to argue for a complementary role for mutual information. Consider the following:

A honey bee returns to its hive after locating a source of nectar. It dances a characteristic dance; its hivemates interpret (in bee fashion) this to mean that there is nectar 100 yards due north. Do the hivemates in bee fashion *know* that the nectar is 100 yards due north? Supposing that the scout did, in fact, just return from a nectar source 100 yards due north and that it dances the correct dance, then the ecumenical naturalist is inclined to say that, yes, the bees do know that there is nectar there. However, if the scout actually just returned from 25 yards due east and accidentally danced the 100-yards-north dance, the fact that there does happen to be nectar 100 yards north and that they are inclined to search for nectar there on the basis of correctly interpreting a well-formed signal is not enough to give the bees knowledge. The signal-foraging response is "true" in bee fashion, but only accidentally true. Something beyond the right signal and the positive payoff is required, even for animal knowledge.

This is the kind of example that is used in old-fashioned epistemology to demonstrate that knowledge must consist in true beliefs that are *justified*. The issue for bee knowledge that corresponds to "justification" in this kind of example is more aptly characterized as reliability. So, just as we ordinarily distinguish truth from justification, and both from utility, here we must distinguish the parallel components. It is one thing for signals to have "biologically normative" meaning (cf. Truth), another for them to pay off in

Information Theory

Table 4.1. *State/Response Joint Probabilities*

Probabilities	Danger (D)	No Danger (¬D)
Slap (S)	.08	.45
No slap (¬S)	.02	.45

reproductive success in a particular instance (cf. Utility), and still another for them to track the relevant states of the environment efficiently (reliability). One cannot simply attempt to reduce all questions of efficiency to payoffs because, depending on the reliability of one's cues and the structure of the payoff structure, it may pay to ignore information-rich signals. The question then is: *just what is it that is being ignored?*

Recall the beavers and their danger signals. Suppose that beavers are a skittish lot so that the joint probability distribution for tail-slapping and genuine danger are as in Table 4.1.

In this case, the environment relevant to the signaling system generates .47 bit of information, with .025 bit of mutual information between the slaps and the environmental states. With payoffs as in Table 4.2, which might be reflective of highly efficient local predators, then it is best to heed the warning.

With payoffs as in Table 4.3, however, which might characterize inefficient predators, it is best to ignore the signal and take your chances.

The information generated by the signal, being a property of the probabilities alone, is the same in both cases. However, whether heeding the signal (and thus continuing to send the signal) is adaptive depends on the precise payoffs for the joint states in addition to the probability distribution of signals and states.

Godfrey-Smith (1991, 1996) used signal detection theory to extend this kind of analysis to situations in which, instead of having a simple binary signal or response, the signal comes in a range of strengths. For each level of signal strength, the probability of each environmental state is given, and the problem is what the *optimal* value of the signal is above which the response is initiated. The lesson is similar in that, depending on the payoffs, different thresholds might be optimal. Moreover, having determined a threshold as

Table 4.2. *Payoffs for Tail-Slapping #1*

Payoffs1	Danger (D)	No Danger (¬D)
Run (R)	−1	−1
Don't Run (¬R)	−100	1

Table 4.3. *Payoffs for Tail-Slapping #2*

Payoffs2	Danger (D)	No Danger (¬D)
Run (R)	−1	−1
Don't Run (¬R)	−2	1

optimal, one is effectively put back in the situation of having a binary response. He assesses the resulting response terms of "Cartesian" ($\Pr(S \mid R)$) and "Jamesian" ($\Pr(R \mid S)$) reliability, and it turns out that which commodity is greater at equilibrium depends critically on the payoffs and signal characteristics, much like in the example of beaver-tail splashes. The distinction between the two kinds of reliability is interesting because it addresses certain common biases in the approach to the characterization of knowledge. Because of the very contentiousness of prioritizing one kind of reliability over the other, however, it seems unlikely that either one can serve as a general measure of reliability.

Mutual information, on the other hand, is symmetrical with respect to these two kinds of reliability (recall the alternative formulations via Bayes's rule), has the advantage of additivity over the combination of independent subsystems, and can be used to characterize system-level reliability as well as the reliability of individual signals. There is a reason, however, for not characterizing mutual information as a measure of reliability. Ordinarily, we intend reliability to be reliability with respect to some function. Mutual information is independent of both payoffs and profitable response arrangements. One can have perfect information via perfect *mis*-correlation of a response mechanism with environmental states. Thus, it is more appropriate to characterize information as a measure of tracking efficiency, rather than of reliability. On the other hand, there is reason to think that in many cases mutual information can be a workable measure of reliability.

INFORMATION AND PAYOFFS

However appealing entropy and mutual information are as ways of characterizing the information richness of an environment and the efficiency with which organisms track that information, if there is no relationship between information and payoffs in terms of Darwinian fitness, there is no reason to think that information is an apt measure of some commodity that evolving knowledge systems might attempt to exploit. Just what is the relationship

Table 4.4. *Canonical Payoffs*

Payoffs C	State 1	State 2	State 3	State 4
Response 1	1	0	0	0
Response 2	0	1	0	0
Response 3	0	0	1	0
Response 4	0	0	0	1

between information and payoffs? For reasons having to do with complexity of mutual information, it is easiest to explore the relationship computationally.

Let us take as canonical the following payoff matrix (Table 4.4) for simple systems for up to four states and responses:

Beginning with a system with two states of the environment and two response states, we can generate probability distributions at random and plot the information of the resulting states against the attendant payoffs. Figure 4.2 gives the results.

What is clear is that high information does not guarantee high payoffs. In fact, as the concentration of points in the lower right-hand corner of the graph show, perfect information can be characteristic of a probability distribution that gets no payoff at all. The reason for this is that the information measure

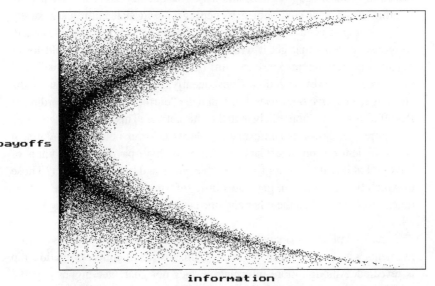

Figure 4.2. Information versus payoffs for two states, maladaptive responses included. Information ranges from 0 (*left*) to 1 bit (*right*); payoffs range from 0 (*bottom*) to 1 (*top*).

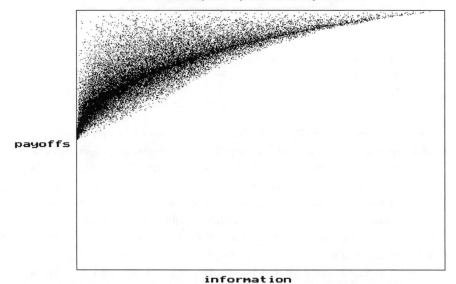

Figure 4.3. Information versus payoffs for two states, maladaptive responses eliminated. Axes are as in Figure 4.2.

does not presuppose any particular mapping from responses to states; it is not biased in favor of adaptive correlations. Consequently, you can have perfect information about an environment by getting it *wrong* all the time just as surely as by getting it right. Natural selection, however, is not so open-minded. Response systems that get it wrong more than right, and thus yield lower payoffs than random response systems (halfway up on Figure 4.2), will tend to be weeded out by selection. Consequently, it makes sense to restrict the analysis to *adaptive responses,* for which the "correct" response according to PayoffsC is always more likely than the alternatives. This gives us Figure 4.3.

Figure 4.3 shows that information places a lower bound on payoffs; it is a sufficient but not necessary condition for high payoffs. What kinds of joint probability distributions yield high payoffs and low information? Those in which the environment generates little information, in which one state is much more likely than the other and one can thus do well without tracking at all.

Figure 4.4 plots information against payoffs for adaptive responses to environments with two equiprobable states. Thus restricted, a clearer relationship is revealed. Although more information does not guarantee higher payoffs, even for our canonical payoff matrix, when we ignore configurations that would be eliminated by selection and focus on single environments (e.g.,

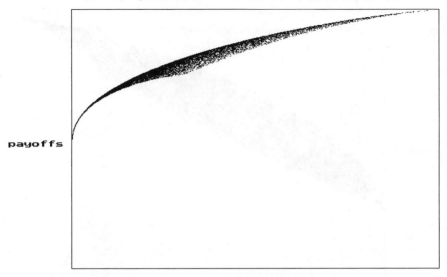

Figure 4.4. Information versus payoffs for two equiprobable states, adaptive responses. Axes are as in Figure 4.2.

Figure 4.5. Information versus payoffs for three equiprobable states, adaptive responses. Axes are as in Figure 4.2.

Figure 4.6. Information versus payoffs for four equiprobable states, adaptive responses. Axes are as in Figure 4.2.

those with equiprobable states), it becomes clear that information places both lower and upper bounds on payoffs. Figures 4.5 and 4.6 show the scatter plots for systems with three and four states or responses, restricted to adaptive responses and equiprobable states. Information still places upper and lower bounds on payoffs.

PARETO OPTIMIZATION OF ADAPTED RESPONSES

There is another way of looking at the relationship between information and payoffs. Instead of thinking of the relationship between payoffs and information in randomly chosen probability distributions, we can focus on the relationship between payoffs and information increase in the process of optimizing response mechanisms. What happens to the information in a response system when its contribution to reproductive success increases? The scatter plots from the previous section indicate the general answer. Increase payoffs and the information can either go up or down, depending on the trade-offs that are made. However, if we restrict the way in which payoffs increase, the result is interesting.

Information Theory

Evolution is an optimizing process in the sense that systems with higher payoffs get selected, no matter what sacrifices might have been made in terms of other kinds of functionality. Presumably, bipedalism brings with it net advantages, at the cost of increases in backaches. Flexible response systems may increase net payoffs while decreasing information, by decreasing the likelihood of expensive miscorrelations via some mechanism that happens to increase the likelihood of cheap miscorrelations. I want to consider system optimization for cases in which trade-offs are ruled out. In economics, Pareto optimization is a process by which the utility for some is increased while the utility for none is decreased, which is to say the whole is optimized without sacrificing the optimization of the parts. I borrow the term and concept to limit optimizations to those that do not exploit trade-offs. Recall that adapted responses as defined previously are those for which the right response is always more common than any of the wrong ones, given the state of the environment. Let the Pareto optimization of an adapted response be changes in the probability distribution which never increase the joint probability of a response with the wrong state and never decrease the joint probability of a response with the right state.

If we Pareto-improve adapted responses generated at random by increasing the likelihood of some correct response by some small δ while decreasing the likelihood of the same response in some other state by the same amount (this leaves the probabilities of the environmental states alone, as it should), we find that for $\mathbf{n} = 2$, information always increases.

Recall that one formulation of the information relationship was

$$I(\mathbf{S};\mathbf{R}) = H(\mathbf{S}) + H(\mathbf{R}) - H(\mathbf{S} \ \& \ \mathbf{R}).$$

Optimization of a response to a state should not change the probabilities of the environmental states. Consequently, $H(\mathbf{S})$ will not be changed by the optimization process. Because of a basic property of entropy shown by Shannon (1949, 51f), the Pareto changes to the probability distribution outlined earlier necessarily reduce $H(\mathbf{S} \ \& \ \mathbf{R})$. Such changes, however, can also reduce $H(\mathbf{R})$ if the result of the optimization is to reduce the probability difference between the two responses. Whether information is increased or decreased, then, depends only on whether $H(\mathbf{S} \ \& \ \mathbf{R})$ decreases more than $H(\mathbf{R})$. By way of example, Figure 4.7 shows that for $\delta = .01$, and $\Pr(R_2 \mid S_2) = .55$ before the modification, the decrease in $H(\mathbf{R})$ is always greater (more positive) than that in $H(\mathbf{S} \ \& \ \mathbf{R})$, no matter the initial probabilities of R_1 and R_2.

Figure 4.8 shows the net effect of the Pareto modification on information across the range of initial probabilities for the adapted response. The relationship holds for all other initial values of $\Pr(R_2 \mid S_2)$ and δ.[11]

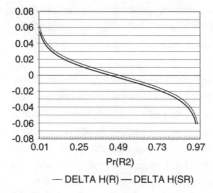

Figure 4.7. Relative change in information components.

No proof of the increase in information on Pareto optimization of responses in systems with **n** > 2 will be forthcoming, for the simple reason that the relationship fails. Although there are counterexamples, however, computer simulations show that failures of information increase in the Pareto optimization of adapted responses occur only rarely in randomly generated probability distributions. For **n** = 3, failures occurred less than 7 times in 1,000; for **n** = 4, less than 4 in 1,000; and for **n** = 5, less than 2 in 1,000.[12]

The conceptual independence of information from payoffs makes it an attractive way of measuring tracking efficiency. Without some strong relationship between information and payoffs, however, there is little reason to think that information is a commodity that might be systematically increased by evolving knowledge systems. What these results indicate is that in systems where response to environmental states are adaptive, information will

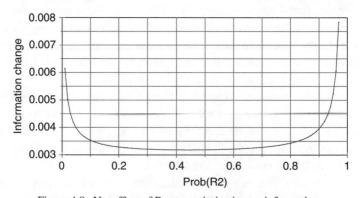

Figure 4.8. Net effect of Pareto optimization on information.

Information Theory

tend to be systematically, if not monotonically, increased as payoffs due to differentiating responses increase.

CONCLUSION: FOUR CONCEPTS OF INFORMATION

Originally, it seems, to "inform" someone was to mould or give form to their mind – to teach. Information, then, would be what you had after you had been informed. Our modern ordinary language concept of information still seems to bear traces of this sort of image (that there is somehow information in a shape) as well as of the correlations of information theory (quantities of information) and information-processing (copying pieces of information). The fact that a term of ordinary language might be equivocal in this way is no cause for alarm, perhaps. What is scandalous is the fact that so many scientists and philosophers, in the midst of trying to be rigorous, help themselves to the concept of information in explaining success while trying to avoid more obviously loaded terms such as "knowledge." It may be that such appeals to information are thought to be justified via Shannon's mathematical theory of information, which is, of course, quite respectable. Much of my purpose in this chapter has been to show what information theory is and is not, from the point of view of prospective uses of information theory by philosophers – what exactly it is that we can help ourselves to in good conscience as things currently stand.

Information theory offers us a measure of correlation or covariance between the states of two systems according to their marginal and joint probabilities (e.g., rather than according to some intrinsic properties of objects involved). It has the advantage of being additive for independent subsystems and is sensitive not only to how well one system tracks another, but also how difficult that tracking is to accomplish. Moreover, it is mathematically "nice" in being continuous and continuously differentiable, and it is backed up by a large body of respectable theory, although most of that body of work will be of little use to philosophers since our primary interests are analytical rather than directed at engineering optimization. It is not in any simple way connected to meaning, nor do relative payoffs for various combinations of states occur in it. This means that we can compare tracking efficiency between systems with radically different payoff structures and ask how payoff structures affect tracking efficiency. The preliminary investigations conducted here indicate that under certain conditions, information places both upper and lower bounds on payoffs. Moreover, what I termed Pareto optimization of adapted responses increases information with high likelihood – 100 percent when there are two responses to two environmental states.[13]

What information theory is not, from the point of view of epistemology, may be usefully characterized in terms of a number of distinct concepts of information, some bound up with everyday usage. There are four I would like to discuss: The *information that, structural information, instructional information,* and *mutual information*.

Dretske thought that he could use information theory to ground semantics via the notion of the *information that*. Intuitively, it seems that the doorbell ringing conveys the information *that* someone has just pushed the button, or at least the information that the button has been depressed. Information theory does not offer, as it stands, an account of this notion of information, for the simple reason that its notion of information admits of infinite degrees, rather than of the binary yes-no of the information *that* concept. (This is not to say that mutual information, or, more generally, conditional probability, plays no role in the correct analysis of information *that,* but merely that such an analysis requires considerably more than what information theory has to offer.) The question it asks is always "how much information" rather than "which information." It trades in a continuous commodity rather than a discrete one.

This point becomes clearer with an example: say, in a human population we discover the sickle-cell gene at a stable 40 percent relative to its normal hemoglobin competitors. We might think that this fact conveys the information that the population confronts malaria on a regular basis. Perhaps it does convey this information *that,* but this is not what information theory can account for, for the theory alone does not have the resources to define the information *that*. What it can account for is the amount of information the population's state contains about the environment. Let S_1 be nonmalarial environments and S_2 be environments with malaria. Let R_1 be populations with more than, say, 20 percent sickle-cell genes, and R_2 be populations with less. We could analyze all populations (real, counterfactual, or both) and assign joint probabilities to the combined system. We could then assess how much information on average the population distribution of sickle-cell versus normal hemoglobin carries about the relevant environmental state, and we can extract the single term for the current state and current response, and say how much information some particular population has about its current environment. In the case of a 2×2 system, the response will only carry positive information about one of the environmental states, but this is compatible with the "right" response being only slightly more likely. Information theory alone assigns no reliability thresholds.

Information Theory

Gene pool examples are nice because all the kinds of information are present. Consider the occurrence of the sickle-cell allele in one individual. The presence of the allele indicates something about how likely the various states of the environment are, and this kind of probabilistic relationship is what information theory deals with.

Sometimes, however, we talk as though there is information in the allele itself – not because of its probable co-occurrences with other things but because of its structure. The allele is a section of a chromosome, which consists of a particular sequence of nucleotides. There are a certain number of possibilities at each point in the string (four), and there are a certain number of points along the string. There is information in this structure, in some sense, although it seems to me that we have no general theory of this kind of information, nor any clear idea of what we mean by it. This is the sense of information that is often intended in the phrase "information processing." (I'll dub this "structural information.") Notice that this is related to the structure of "information spaces" that play a role in the optimization theorems of information theory. For cases in which there are discrete alphabets that make up characteristic "messages," one can envision a general theory of structural information, but it is hard to see (at least for me now) how this might extend to all the cases in which we are inclined to say that there is structural information, especially ones for which we know of no basic "alphabet."

We also sometimes talk about something one might call "instructional information," information "how to." The allele contains, in the context of the decoding scheme for protein synthesis, the information about how to deform blood cells resulting in the mixed benefits of sickle-cell anemia. Presumably, in the context of a different translation scheme, the allele would contain the instructions for how to make something else, or, more likely, nothing at all.

Notice that these concepts of information are distinct. The occurrence of the allele with a certain probability reflects something about the environment. How much it reflects depends first on how much information the environment generates (i.e., how likely are high rates of malaria) and how well correlated the occurrences of the allele are with environmental states. This is mutual information. It is distinct from the information in the structure of the gene, and both are distinct from the biologically normative causal consequence of the occurrence of that allele. What the three have in common is the occurrence of a certain token of the sickle-cell allele type, but the three kinds of information arise from different aspects of that occurrence. Mutual information arises from the probability of the occurrence in context. The particular structure only serves to identify what *kind* of occurrence it is, but aside from determining identity, the structure is irrelevant to mutual information. Structural information (whatever that might be) does not seem to depend on the probability of the occurrence of a type, but only on the details of its construction. Instructional information does not depend on probabilities, but it does depend on structure. *How* it depends on structure, however, is highly sensitive to context. Indeed, whereas mutual information and structural information may be thought of as purely descriptive concepts, instructional information has strong normative overtones, as does the information *that*. Presumably, these latter concepts of information can be cashed out in terms of biological functions (see Part III).

The role I suggest for information theory in naturalistic epistemology is, in some ways, quite modest. It is not a panacea. It does not reveal deep truths about the nature of the universe. It does not even suggest a metaphysics. It does not account for the variety of meanings of "information" in everyday language. It does not play a decisive role in the theory of meaning. What it does offer us is a measure of tracking efficiency that is indifferent to ontologies, independent of payoffs, and mathematically elegant. This turns out to be exactly the tool we need for evolutionary epistemology.[14]

5

Selection as an Information-Transfer Process

The purpose of this chapter is, finally, to begin to demonstrate the significance of the concepts of evolution and information developed in the previous two chapters for the project of naturalizing epistemology. The development of the information-transfer model that begins in this chapter and proceeds over the next two becomes increasingly technical because, unlike the previous two chapters, it is not an introduction to a well-established tool set but the application of those tools to the problems of epistemology. For those more interested in the philosophical payoff than the details of the model itself, I try to summarize its importance at the end of Chapter 7. Before proceeding with the development of the model, however, it is important to remind various kinds of readers what it is that is being attempted and what we may expect to accomplish. We have, as it were, arrived at the foot of our chosen mountain, newly equipped, and the time has come to focus our thoughts on the challenge ahead.

For those who have not been thinking about epistemology lately (we have certainly not been thinking about it much in the last few chapters), a brief reminder of the nature of the problem may be helpful. For those who have been thinking about epistemology, the manner of my presentation of the problem will tell them something about where I stand on the philosophical landscape, although I address that directly in due course.

As I see things, what has come down to us from a long and varied philosophical history, which traces back at least to Plato and bursts into full bloom in the seventeenth and eighteenth centuries, is the following: Concepts such as "knowledge" and "truth" have always been central to our understanding of the factual inquiry which eventually grew into contemporary science and have always been puzzling in a way that threatens to undermine the credibility of anyone who claims to be capturing what reality is really like in their theories and observations. The fundamental dichotomy is the old one between appearance and reality, and the fundamental problem is how to say anything non-question begging about the relationship between the two. The problem was evident in the concerns of the ancient sceptics and in the rationalists, and the issue was finally forced on the philosophical tradition via the empiricist model of perception developed by Locke and deployed devastatingly by Hume. Kant, one of the finest technical minds in the Western tradition, announced to the world in his *Critique of Pure Reason* the impact of Hume's skeptical empiricism and the appearance-reality distinction took in his hands perhaps its starkest form. Kant contrasted *phenomena* – the world of everything we experience directly – with *noumena*, the merely hypothesized and wholly unknown things-in-themselves, what Hume referred to metaphorically as the "hidden springs and principles of nature."

To make the problem as clear as possible, consider what would be involved in comparing the appearances you experience to the reality you presume those appearances to represent.[1] At least in the most obvious attempt, what you would have to be able to do is somehow access the reality independently of appearances, but this is impossible. You can never inspect or experience and object "directly" and then compare that experience to the appearance you perceive. You can, of course, consult other appearances of the presumptively *same* thing and compare those to the first. Turn the apple and look at it from another angle. Ask another observer for her opinion on its color. Touch and smell the apple along with looking at it. This is, of course, what we do, and the utility of this kind of triangulation on the world via multiple observations, multiple sensory modalities, and via the experiences of multiple individuals is the source of what objectivity we have. Nonetheless, the Kantian barrier to reality as it exists in itself remains. This is evident in the puzzling question of what it could mean for our thoughts to resemble reality. It is evident in the current difficulty with coming up with a defensible naturalistic account of the intentionality of thoughts and language. And the presumptive status of human reason as a product of evolutionary transformation even undermines Kantian attempts at internal or a priori proofs of necessary regularities of all experience.

Contemporary epistemologists insist that, if the skeptic takes the undeniability of the Kantian barrier to mean that knowledge is impossible, she is applying impossible and unrealistic standards to knowledge. Whatever knowledge is, it need not require that we somehow see reality as it ultimately is, if that even makes sense. So knowledge, while perhaps less final than some philosophers have wished, is nonetheless both valuable and attainable. Fair enough. But the Kantian barrier is still with us, for the mere recognition that we have never been able to inspect reality directly (and thus *that* cannot be what the word *knowledge* refers to) does nothing to explain exactly what the nature of the relationship is.

One might ask why we need such an explanation. Living with the realization of the epistemological barrier I take to be the very cornerstone of philosophical skepticism – restraining in oneself the urge to make unwarranted assumptions about the nature of reality in order to create the sense of security that knowledge provides. That was, after all, what the *epoche* or "suspension of judgment" was all about. Most scientists and, indeed, people of good sense in general, do not and probably should not consider the Kantian barrier to be an impediment to their work or to living productive and happy lives. But then sometimes the generation of reasoned consensus is sabotaged by the available option of relativism – perhaps that is true *for you*. Or your ability to "win" this argument merely reflects that you stand with the dominant paradigm. Or you are simply bullying me into agreement. At such points, it would be nice to be able to say not so much what *the* truth is, but at least what *truth* is, what it means to know something, and thus what our best strategy is for coming to agreement.

What this project is about is not eliminating or overcoming the Kantian barrier, but rather attempting to fulfill the suggestions of many students of knowledge such as Campbell, Dewey, and Quine to the effect that the principle of natural selection can somehow be used to put a satisfying dent in the epistemological "veil of ignorance." The fundamental intuition behind the current project should be familiar. While we may be able to say little about the natural divisions or ultimate nature of reality (at least to start with), what we can say is that insofar as it makes itself known to us, it does so through patterns and regularities in our experience. Obviously, not all such regularities reflect the structure of the external world, but many of the regularities we experience seem unaccountable in terms of internal constraints. Counterintuitive phenomena such as the angular momentum of gyroscopes are a good example. So, reality does affect us, and if we simply focus on patterns of experience without making unwarranted assumptions about their causes, we can begin to get a handle on the problem of knowledge. This intuition is, of course, much of what lies behind "sense-data" empiricist approaches to knowledge such as Locke's and early-twentieth-century logical positivism. One also sees it in a more sophisticated form in Dretske's work (e.g., 1986). This focus on the patterns of perceptual input leaves out much of the story, however, for it does not help us understand the way we respond to external stimuli nor why the character of our responses should be in any way beneficial. What we need is to recognize that the world not only stimulates but *selects* as well. It selects individuals whose response patterns interpret stimuli in a beneficial way. Understanding what it is in the world we are responding to is at least as much a matter of understanding the ways in which the world selects cognitive mechanisms as it is of examining the behavior of those mechanisms. For the patterns of our experience can only suggest that our synchronized and confirmed perceptions are tracking *something* in the world, but not what is being tracked or even that our conception of the source behind our synchronized experiences are systematically useful representations of those sources. Selection and stimulation need to be modeled in parallel, rather than insisting on the primary importance of one and rejecting the other because it too severely underdetermines the final behavior of organisms like ourselves.

At this point, it should become clear why I have developed versions of evolutionary theory and information theory which are as ontologically neutral as possible. If natural selection is only something that happens to entities as ontologically specific as a replicator (they are general but not that general), then we cannot say much about how they reflect the nature of the world without making some fairly strong assumptions about the nature of the world. If, however, natural selection is something that happens to any typed collection of entities, and, if furthermore, the nature of the world itself is what determines the (expected and probable) fitness of whichever typed collection interests us, then we can get by with the very minimal assumption that local reality lies behind the observed or inferred differential stability tendencies of the various types. Similarly, information as we have construed it is also a fully general concept which brings with it no ontological assumptions, but only the requirement that we be able somehow to specify world-states and receiver-states. Of course, we can no more measure information of this sort in our actual situation than we can step outside of our relationship to reality and objectively compare the nature of our thoughts to the things they are thoughts of. Still, we will see that an ontologically neutral measure of tracking efficiency such as mutual information allows us to think about different ways in which the relationship between our cognitive system and the rest of the world (for we are, after all, part of the world) may be systematically tightened by the simple stability requirements which affect all distributions of entities.

What makes the principle of natural selection as developed in Chapter 3 epistemologically interesting is the very generality and ontological neutrality that tempts critics to accuse it of tautology or circularity. Even in the absence of any opinion on the ultimate nature of things, that the stability of various sorts of (for the lack of a better word) "things" differ in and across environments because of interactions with those environments amounts to little more than the virtually unavoidable assumption that there are causal interactions between things that affect their persistence, where those so-called things are in their own right just whatever local configurations of world-stuff are accountable for the otherwise improbable observational agreements across time and between persons. So if we make the tiny little "leap of trust" that is necessary (I won't call it faith for faith seems to require a certain tenacity in the face of *contrary* evidence), we should expect a certain local correlation between the occurrent properties of things and their utility vis-à-vis our continued persistence relative to the local environment. Or the mere fact of selection entails that variability tracks local environments in useful ways. This basic cornerstone of evolutionary epistemology may fall short of conceptual necessity (and even if it didn't, I must insist that conceptual necessity reveals much about the nature of our system of representations and only obliquely gets at what the world would have to be like for such a system to be useful). Here, however, the name of the game is not foundationalism's something for nothing, but rather science's a lot for a little. What makes it epistemologically interesting is that we can start to say something about how complex systems like us might relate to the world merely on the basis of the fact of our *persistence* in a varying environment.[2] One part of this is the implication that our various thinking processes are tracking environmental change in useful and reliable ways. The more interesting bit for me is that we can understand our conceptual categories for states of the world in terms of behavioral differences that have practical consequences, like the beavers in the last chapter who, via the utility of their tail slaps and avoidance behavior, divided world-states into dangerous and nondangerous. In Part III of this book, we consider what it would mean for selective histories to ground conventions or rules for the correct functioning of our shared representational systems.

So my seemingly perverse celebration of the facts that natural selection happens to just about everything (Chapter 3) and that there is really no such thing as information (Chapter 4) is explained by the fact that such ontologically neutral and at the same time rigorous analytical frameworks are the very best sorts of tools we could hope for if we are interested in exploring not beyond the Kantian barrier, but the nature of the barrier and our relationship to what lies beyond. The minimalist concepts of evolution and information allow us to study the nature of knowledge, not without assumption but with the sort of minimal and productive assumptions that may be hoped to generate a theory of knowledge that is both scientifically productive and philosophically respectable.

Finally, a note on what we are after here. My own affinity for the skepticism of Hume, and indeed for that of the Pyrrhonians whom Hume failed to recognize as his true intellectual predecessors, should be clear by now. Skepticism as I understand it is not based on any positive thesis, but consists of a disciplined restraint on metaphysical reasoning driven by the personal frustration with fine-grained arguments about the necessary nature of that which is evidently beyond our ability to perceive. The skeptic suspects that metaphysical arguments which appear to establish necessary features of the reality in fact end up doing far less than that and trace, at best, the limitations of our own thinking. What is right about phenomenological approaches is the seriousness with which they take the Kantian barrier. What is wrong with them is that they let the desire for certainty or incorrigibility determine the nature of their inquiry, just as Descartes's desire for certainty led him to consider solipsism as an option, if only temporarily. Thus, skepticism for me results in a kind of pragmatism – if one is going to theorize, one might as well try to build something useful, or at least the kind of thing that tends to be useful. Pragmatic considerations direct one toward materialism, taking successful science as a model. Foundationalist-reductionist frameworks recommend themselves not because the ambitions of foundationalism (attaining the kind of certainty geometry brings) are any longer plausible, but because reductionist frameworks facilitate the maximum integration of various bodies of knowledge. The task, then, is to begin the construction of an analytic framework which maintains maximum ontological neutrality via the use of mathematical concepts and yet allows us to say interesting (and hopefully useful) things about the relationship of our thoughts to the world.

PUTTING IT ALL TOGETHER

At this point, we need to go back and consider our basic evolutionary dynamics under selection alone from the epistemological point of view. Figure 5.1 shows a population "learning" about its environment.

The idea is that you take some collection of objects, organisms, or what have you and categorize them according to some type scheme. (The type scheme is our contribution to the model, from the phenomenological side of things.) Figure 5.1 shows the a population of sixteen types with equal initial frequencies, but neither the number of types nor the initial frequencies are critical. The line graph above the bar graph gives the absolute stabilities (fitnesses) for the sixteen types which, since the fitnesses are all greater than one, implies that all sixteen types are either reproducing or propagating via imitation or some similar process. Over the course of four hundred cycles, the population becomes dominated by the most fit type. This result is inevitable in the simple case with fixed fitnesses and no variation.

Selection as an Information-Transfer Process

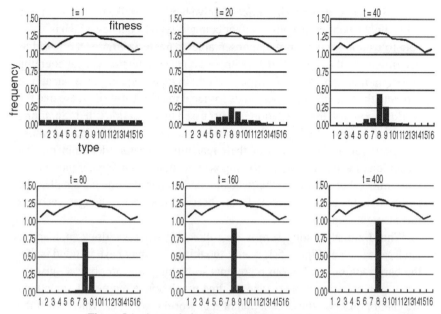

Figure 5.1. A population learning about its environment.

In light of the preceding discussion regarding the basic problem of epistemology, several features of Figure 5.1 should stand out. The first is that we have *imposed* our type scheme on the population via the process of counting how many individuals there are of each type. We do not know whether we have categorized the members of the population correctly or, for that matter, what it would mean for us to categorize them correctly. What matters methodologically is that the categorizing and counting is reliable and repeatable and that the categories are good for the purposes of gathering scientific knowledge. (The fact of reliability-repeatablility indicates an exploitable relationship between us and the real source of observational stimuli. Scientifically useful categories require clear logical and operational consequences and coherence with the rest of the body of knowledge.) For current purposes, however, it really doesn't matter what categories we use. The important thing is that when we counted and categorized, *the world* contributed to the stability properties for each type so constructed. Which is to say that whereas the characterization of the population reflects our thinking and perception, the fitness vector for *that* population is automatically given by the world. This is not to say that the fitnesses are measurable, for unlike the simulations depicted in Figure 5.1 in which the fitnesses are stipulated, fitnesses in the real world cannot be measured directly but only assessed by observing the evolution of the population itself. So what is depicted in *model* adaptive landscapes like the one in Figure 5.1 is, on the bottom, the observed- or counted-type frequency which is available to us and on the top the unobservable characterization of the world as determined by the type scheme we impose on the population via our counting and categorization. When we observe shifts in type frequency, we can at best estimate the true fitness or potentials for local multiplicative increase that the world imposes on the population. As discussed in the section titled "Sampling Error" at the end of Chapter 3, our models assume large populations whose evolutions obey the dictates of their real fitness vectors deterministically. Evolution of actual populations composed of smaller numbers of individuals only reflect that vector probabilistically, and as populations get smaller, the ability to infer actual fitnesses from observed frequency changes becomes increasingly impaired.

What I am suggesting here is the following: when we draw an adaptive landscape such as Figure 5.1, one of the things we may have in mind is that the bar graph on the bottom represents something we can observe, just the result of a counting process, and the line graph above represents something we cannot observe, the summation of real-world tendencies which affect the stability and proliferation of types. Thus, adaptive landscape as construed in terms of observable frequencies of any arbitrary population and its real fitness vector gives us a sort of model of the Kantian barrier, for the determinates of real fitness lie beyond that barrier, but its effects on observable populations lie on this one. This construal of the adaptive landscape will be the basic model we will develop for how information is transferred from the world into populations, and that development will be a matter of adding other interacting populations with their own type-scheme-driven construal of the external world. The model of our own version of the Kantian barrier will, of course, be much more complex (and ultimately beyond the scope of this book) but will depend on the same principled characterization in terms of population type-schemes and the fitnesses that the world provides for them. In the meantime, Figure 5.1 characterizes the problem of knowledge as it confronts, for instance, the pots in my garden, rather than any organism complex enough to represent its environment. In such a "zero-case" of knowledge, we find in its simplest form the principles of stability-driven information transfer which manifest themselves in more complex ways in our own system of knowledge.

So, first, the fitness vector as it appears in the adaptive landscape is understood as being a representation of forces which lie beyond the Kantian barrier. This by itself is nothing special, since we firmly hope that many of our theories similarly represent the real world. What is epistemologically significant about the real fitness vector is that the world determines its value for any arbitrarily chosen population. This will give us a way of thinking sytematically about how other and, indeed, any population slices its world. We, of course, have our own concept set, and we surely have no choice but to employ it when we study knowledge, or anything else for that matter. But what we do not have to do is insist that all systems are to be judged by the extent to which they have attained our own ideal representation of the world. Ultimately, this will allow us to account for our own concept set in ways that do not begin by presuming its own legitimacy.

Second, we have already seen one way in which the information a population may display about its fitness vector is limited in that finite populations only obey its determinations probabilistically. In addition, much of the detailed structure of $\vec{w}$ is, even in the best cases, only reflected in nuances in the transitory evolution of the population. In Figure 5.1, the fact that type 2's fitness is higher than that of types 1 and 3, for instance, is only briefly reflected in type 2's temporary advantage over its neighbors (at $t = 20$) but is washed out in the long-term dynamics whereby the most fit type goes to fixation. Thus, there is in some sense far more information in $\vec{w}$ than the population is capable of gathering. In the simplest environments with fixed fitnesses, selection does lead reliably to the population "identifying" the most fit type, but other features of $\vec{w}$, such as the relative fitness rankings of the various types, are lost. This particular limitation is not due to the Kantian barrier so much as to the nature of the "learning algorithm" itself – natural selection. What, then, of the rest of the vector $\vec{w}$? The significance of $\vec{w}$ as an epistemic target is that $\vec{w}$ and its own evolution over time (remember that fitnesses typically are not fixed and are frequently dependent on a population's own frequency distribution) and space constitutes absolutely everything a population would ever find useful to track – in selection-driven systems, everything it could possibly track. In some sense, $\vec{w}$ just is the world as it exists, not in representation, but in practical potential for a given population. If this is right, then we can understand the increasing complexity and proliferation of distinguishable world-states as driven by increasing complexity and proliferation of the type schemes of populations. Here, the full generality of our evolutionary model holds its greatest promise, for its ability to accommodate populations of beliefs means we will be able to understand, in the same terms, the way states (and ultimately objects) are distinguished for those populations.

The critical point for the immediate development of the model is that on any type scheme, selection causes populations to evolve such as to reveal something about the local environment as reflected in the fitness vector for that type scheme. What is revealed is just which type is most fit, with the finer nuances of the comparative fitnesses washed out by the eventual elimination of all but the most fit type. Imagine, for instance, that the line graphs indicating fitnesses were eliminated from Figure 5.1. One could tell simply by looking at the population's evolution which type was most fit, at least on the assumptions of fixed fitness and no variation.

It appears, then, that the evolution of a population indicates something about the environment as characterized by the population's fitness vector. In simple cases of fixed fitness and no variation, large populations unerringly pick out the most fit type. We focus on this simplest of cases here, but it is also the case that the evolution of populations with frequency-dependent fitnesses indicates something about the environment as well, if only that the environment is such that which type is most fit is a function of the population distribution itself. We will have our work cut out for us even taking the simple case of fixed fitness beyond the mere recognition that populations indicate something about their environment to a more precise specification of what it would mean for information about the environment to be gained by the population under selection.

STATES OF THE WORLD AND THE POPULATION

The graphical results in the last chapter demonstrated certain relationships between information and utility (or fitness) for simple cases with small numbers of environmental states and adapted responses. Information places upper and lower bounds on utility and, in the case of two-state, two-response systems, "Pareto"-improvements in the tracking of response systems result in increases in information.

The pioneering mathematical biologist R. A. Fisher proved some time ago a certain relationship between fitness and selection. Under certain assumptions, average fitness increases at each step ("monotonically") under selection, approaching a maximum. Called Fisher's fundamental theorem, this result applies not only to systems with fixed fitnesses, but also to some systems with frequency-dependent fitnesses (i.e., with a fitness matrix). The restriction involved is that the fitness of any strategy i against any other strategy j must be the same as that of j against i. This symmetry requirement on the fitness matrix holds true in general for alleles in a gene pool, where the fates of matched alleles are common because of their shared destiny within the same cells. Game-theoretic situations, such as the Prisoner's Dilemma game, do not in general satisfy the matrix symmetry requirement. Cooperators and defectors do not do equally well when they play against each other, so that Fisher's result is of relatively little interest outside of the realm of

Selection as an Information-Transfer Process

evolutionary genetics. For generalized evolutionary theory, selection commonly drives average fitness *down*.

Our task here is to try to understand how selection affects information over time. Given the generality of the mathematical concepts of evolution and information, it may be that little or nothing can be said about how information evolves for all systems. The strategy instead is to develop highly simplified models that allow us to understand the basic principles of information transfer under selection. Then we can begin to investigate more complex systems – not all such systems, but those with the sort of complexity that one sees emerging as biological cognition becomes more complex.

Applying mutual information to the environment-population relationship requires that we be able to characterize the spaces of the fitness and distribution vectors in terms of states of sending and receiving devices. Note that there are continuous versions of mutual information that could be applied directly to the relationship between the two vectors, but the discrete version is simpler and more appropriate for epistemological applications. Figure 5.1 nicely illustrates the point that for fixed fitnesses the main fact one is likely able to deduce from a population's evolution is which type is locally the most fit. This suggests that we assign states to the environment by imposing a *partition* on the space of the fitness vector. There will be **n** cells in the partition, one for each type, and they will be of equal size and symmetrical. We can define the cells by stipulating that the environment is in state i when the fitness vector is in cell i, which is to say, when type i is the most fit. Strictly speaking, there is also a small region where one or more types share the same highest fitness, but the chance of this being the case is small enough to be ignored.

Let's call the states of the environment $S_1 \ldots S_n$, where S_i is the state where type i is the most fit. This leaves us to decide what the "receiving" states should be. Recall that mutual information does not require that there be the same number of sending as receiving states, although the maximum information is determined by the minimal number of states that sender or receiver have. This gives us a good bit of leeway in how we characterize the receiving system, and it will turn out that there are lots of interesting ways to do this.

In this book, we are characterizing environments in terms of which type is the most fit. There doubtless are other interesting characterizations, but this seems like the right place to start. There are three characterizations of the receiver we use. One of these is used in the formal proof of information transfer in the Appendix, two are used for the computational analysis in Chapter 7.

Given what we have done so far, the obvious thing to do is to partition the distribution vector's space just as we have partitioned the fitness vector's space, so that there are **n** receiver states determined by which component of the distribution vector is the largest. Whereas in the case of the fitness vector, the **n** states were determined by which type was the most fit, the **n** states of the distribution vector are determined by which type is the most *frequent*. So, we characterize an environment-as-sender as being in **S1** when Type 1 has the highest fitness, in **S2** when type 2 has the highest fitness, and so on. Similarly, we characterize the population as being in **R1** when type 1 is the most frequent, in **R2** when type 2 is the most frequent, and so on. In terms of the adaptive landscape representation, environment-sender states are determined by which point on the line graph is highest, population-receiver states by which bar on the bar graph is highest. The partitions are similar except that the distribution vector is restricted to the simplex (the hypertetrahedron in which the **n** components add to one) and the fitness vector is free to range over its whole **n**-dimensional space. This "which type is the most frequent" partition is used in the proof in the Appendix.

THE PROOF

The Appendix contains a formal "proof of information gain in frequency independent replicator dynamics for populations of $\underline{n}$ types." What the proof proves is that for fixed fitnesses and no variation and for the states of environment and population just described, information increases monotonically (at each step) over time under selection (the replicator dynamics). Selection, then, is an information-transfer process, at least insofar as the setup involved in the proof is taken as basic. The reader who is so inclined is invited to work through the proof itself, which I have tried to make more readable than I usually find such things to be.

Since information is a matter of probabilities and since the assessment of how information evolves over time requires, roughly, counting how often world states and population-states occur together as selection progresses, proving anything requires that the probability of various fitness vectors and initial population distributions be specified. The obvious thing to do would be to stipulate that all joint configurations of initial distribution and fitness vectors are equally likely, and the proof works just fine on such an assumption. The requirement that is actually stipulated is a weaker symmetry requirement, which simply says that each of the types is alike with respect to its chances of initial frequency and fitness. As a result, the proof has somewhat broader implications.

The proof itself is not as broad or elegant as I would have liked. I am convinced that more general results are possible, although my own limitations as a

Selection as an Information-Transfer Process

mathematician were challenged enough in the generation of this simple result. In particular, there are without a doubt many other partitions for which selection increases information, and it would be nice to be able to say something about the range of partitions for environment and population for which the result holds. (The fewer assumptions about states of the world, the broader the implications are for information gain under selection.) Likewise, some of the computational results from Chapter 7 seem to indicate that uniform mutation patterns may not compromise environmental tracking at the population level. On the other hand, the symmetry assumption is nicely general, and the assumption of fixed fitness does not seem to me to compromise the import of the result. Variable or frequency-dependent fitnesses mean that selection pressures are changing constantly. What the monotonicity of information gain that follows from the proof indicates is that *at each instant*, while fitnesses are fixed, the force of selection is driving information up. If fitnesses change as a result of population frequencies or external factors, then strictly speaking there is a new environmental state to track. Consequently, the unrealistic characteristic of the assumption of fixed fitnesses is largely mitigated by the monotonicity of the result. On the other hand, variation, which is excluded in the proof, can be expected to counteract the systematic force of information transfer in real systems, as will the sampling error that occurs in the dynamics of finite populations. So, to say that selection is an information-transfer process is not to say that there is no other force opposing or compromising it, but rather that insofar as selection dominates and populations are large, one can expect the systematic transfer of information.

OTHER RECEIVER CHARACTERIZATIONS

Like most proofs in the area of dynamical systems, the assumptions for the proof just discussed were chosen not so much on the basis of what is interesting as on the basis of what will allow the proof to go through. One of the attractions of building computer simulations as opposed to engaging in mathematical proofs is the freedom one has to choose assumptions that are interesting. In this case, the characterization of receiver systems used in Chapter 7 was chosen on the basis of interest rather than analytical tractability. Two rather different characterizations were used.

The first, "population-level information," is not unlike that used in the proof in that it partitions the space of the distribution vector. The difference is in the number of states. In Chapter 7, we consider populations with just two types, which have the feature that the state of the population can be described

Information and Meaning in Evolutionary Processes

with just one number between zero and one; for example, the frequency of type one, since the frequency of the other type is just the remainder. Instead of just two states determined by which type is the most frequent, on this partition there are ten states of the population depending on whether the frequency of type one is between 0 and 0.1, between 0.1 and 0.2, between 0.2 and 0.3, and so forth. This gives us ten states of the population and just two states of the environment, with a maximum of 1 bit of mutual information.

The second characterization, "individual-level information," does not partition the space of the distribution vector at all. Instead, the frequencies of types are used to assess not the probability of the population being in some *state*, but the probability of individuals being of each *type*. In this case, there are again **n** states of the receiving device, but the receiving devices are the individuals, in the following sense. Individuals are not indicating environmental states by responding or changing type, but by *being* of type they are. The receiver states $\mathbf{R}_i$ are the possible types an individual might be, and the probability of $\mathbf{R}_i$ is just the chance of a randomly chosen individual being of type i. The probabilities of an individual being a type change over time, given whichever selection pressures are operative. So, information on the individual level is a matter of using type frequencies directly to assess environmental tracking or information, rather than using them indirectly to first determine the state of the population, where the probabilities of that state are used to assess information. Intuitively, population-level information is a matter of what the state of the population as a whole indicates about the environment. Individual-level information is a matter of what *my* being of *this* type indicates about the environment.

The distinction between these two kinds of information may seem overly subtle at this point, but Chapter 7 will show that there are important differences between the two. In particular, population-level information may be much higher than individual-level information – information that is present in the population but may not be exploitable. Individual-level information seems to have a stronger relationship to fitness than population-level information. On the other hand, if you work through the proof, you will see that a similar technique can be used to prove monotonic increase in individual-level information as well.

THE SLOGAN

Natural selection is an information-transfer process. This is the moral, message, or slogan that one gets from the proof discussed here. Like any slogan, however, it needs to be unpacked to avoid the misunderstandings that are inevitable with simple statements of complex relationships.

In the first place, as should be abundantly clear by now, describing what happens to information as "transfer" or "transmission" is potentially confusing. Information is not a kind of *thing* which is first in one place and then moves to another. It is not even like an idea that can be propagated giving the illusion of a moving thing when the real process is more akin to a chain reaction. Mutual information can exist between two systems with states that bear no resemblance, where there is no sensible way in which one can say that the same "something" that was in the sender is now in the receiver. Information is not sent from the environment to the population via the channel of natural selection, even though as a rough heuristic, that's not a bad way to think about what happens. What really happens is that selection sort of *induces* a probabilistic tracking relationship between population distributions and features of the environment which affect the fitness of the types in the population. Information is the tracking relationship, "transfer" is a metaphor for this induction process.

Thus warned, what makes this information-transfer process epistemologically interesting is that selection transfers information (i.e., maintains the tracking relationship) across a gap that resembles the epistemological gap between our thoughts and the things-in-themselves, for what is on the other side of the gap, on the environment side of the information-transfer process, is presumably just the real causes in the real world that actually are responsible for what happens next in the world. (Notice again that $\bar{w}$ is not the observed fitnesses of the population's types but the vector of probabilities with a value estimated on the basis of observed frequency change.) By being so carefully ontologically neutral, saying nothing about the structure of the environment but merely labeling its tendencies to affect the population distribution, we have avoided the epistemological faux pas of assuming too much about the nature of the world. Moreover, since information does not require a mapping from the states of sender to those of the receiver, mutual information can exist between a population distribution and the natural divisions of the world, if there be such divisions, insofar as those divisions have systematic effects on the evolution of the population. Thus, the combination of mutual information and the mathematical concept of selection gives us the tools for thinking in general about how evolving populations track whatever is going on "out there" in the real world.

The next point has to do with the exploitability of the information relationship. As we saw in Chapter 4, mutual information does have a distinct relationship to utility or fitness, but it is not necessarily the case that high information means high fitness. The distinctive dual-pronged shape of Figure 4.2 shows that the highest information is mathematically consistent with the lowest possible utility, where the tracking relationship consists of always doing (or being) the wrong kind of thing for the state of the world. What selection must do whenever some cost is involved in the response mechanism is weed out the configurations where information tracking results in lower utility that random responses. The result is that, as is shown in Figures 4.3 and 4.4, information places lower bounds on utility or fitness. I describe this effect by saying that selection not only transfers information, but also ensures that the information is *exploitable*. Here also, we must be careful not to misunderstand this as saying that the information is in a representation that can be consulted by a cognitive system. The exploitability of the information just consists in information placing lower bounds on utility, where increases in information raise that lower bound.

We can say, then, that selection can be understood to transfer exploitable information from the local configuration of the "things-in-themselves" to population distributions in general. This does not, however, get us any closer to saying what those things are *really* like, but then, we gave up on that ambition when we ceded victory to the skeptic, along with the hope of any assumption-free reassurance regarding the reliability of our knowledge or the truth of our beliefs. The study of knowledge ceased to be the pursuit of such absolutes and became instead a question of how much, theoretically, we can get for how little. It became a matter of trying to understand within the framework of the natural sciences how cognitive systems track and represent the environment in a usable manner. The skeptic's legacy reminds us that the more assumptions we make about the structure of the environment, the more we are simply asking about how the response system we are trying to understand compares to *us* in the way it individuates environmental states, and the less we will understand about the basic nature of the knowledge relationship itself.

Even as far as establishing the merely *theoretical* foundations for an evolutionary epistemology, establishing selection as an information-transfer process is only the first step. For while we may be satisfied that our characterization of what is on the other side of the information relationship is as close as we are likely to get to being able to talk directly about the real stuff of the world, what is on *this* side of the relationship – the population distribution – is not yet what we are looking for. To be sure, the generality with which the evolutionary process has been defined allows us to have populations of anything at all, which includes distributions of acquired behavioral dispositions like beliefs. The problem arises from the very trick that allows us to get at the world via the mere definition of a population. The world *automatically* contributes the fitness vector for any population we define, and it is information with respect to *that* vector that is transferred. The problem is that the sorts of factors which directly affect the propagation of beliefs are not necessarily the kinds of factors we would like our beliefs to track. Presumably, the immediate reason beliefs propagate is because the belief-environment is such as to cause them to propagate rather than their tracking of the sorts of world-states that have important causal consequences for human well-being. What we need, then, is a model of information transfer in *multitiered* selection processes, where natural selection operates on innate tendencies for behavior and belief acquisition, as well as on the evolution of behavior and beliefs over time. The next two chapters are devoted to the articulation of simple two- and three-tier models of the sort required.

6

Multilevel Information Transfer

Every attempt to bring an understanding of evolutionary history to bear on questions of human behavior eventually runs up against the problem, the undeniable *fact,* of the extreme flexibility of human behavior. More than any other species, it would seem, human beings have the ability to overcome virtually any behavioral tendency with which evolution, in the current guise of "human nature," supplies us. For those who reserve a special place for humanity outside of (or above) the animal kingdom, our ability to learn, to overcome our animal heritage, is proof enough that there is more to us than mere biology can explain; more than for which the nearly three-billion-year history of genetic evolution can account.

Even if we feel it is a mistake to try and understand human beings in isolation, as a species unto itself that follows unique rules and procedures, still one must admit that the flexibility of human nature presents a problem. What exactly is the importance of evolutionary history for a species that can apparently learn (or unlearn) *anything,* other than that history has supplied us with the ability to learn and some tendencies that can be overcome? If our *nature* is such as to make us so nearly entirely creatures of *nurture,* then isn't the essential story about what people actually do a story about our developmental environments, our life histories, our culture, and *its* history?

It may be that, after all, matters of social policy are best conducted on the basis of a certain tenacious agnosticism about innate variance in human abilities and on the basis of the corresponding operational conviction that whatever genetic determinates of human behavior there may be are irrelevant to understanding why people do what they do. It may be that much of the resistance to evolutionary accounts of human nature is due to the impulse to defend the politically very agreeable proposition that all human beings are created equal. Yet, even if the proposition were true as a matter of current biological fact, epistemology still requires both the history of human biological evolution and a model of the *interaction* between evolved human nature and learned human behavior. Biological history is required not only to account for the reliability of our sensory and cognitive apparatus, but is also essential if we are to understand why we construct our subjective worlds the way we do. This is the case even if all humans are currently identical with respect to those abilities. A model of the nature-behavior interaction is necessary to understand how the reliability provided via evolutionary selection by the real world can be exploited by a process capable of open-ended creativity.

The multilevel selection paradigm articulated by Campbell provides exactly what is needed, at least informally. Variation and selection evolutionary processes are ubiquitous, with trial-and-error learning being easily modeled in these terms. The primary way in which the results of biological evolution influence evolutionary learning is not via the generation of trials (although there may be strategies devoted to improving on random generation of trials) but in the detection of error, not via variation but via selection. Creativity is left free to explore whatever bizarre inventions and recombinations of old ideas it can. What *connects* this free invention to the world is the silence of error detectors such as physical pain, hunger, thirst, or cold; the silence of longing, the absence of cognitive dissonance which tells us that something is amiss with our system of representations. These objecting voices of error carry information about the world, they track probabilistically changing features of the things in themselves which are essential to our survival and well-being. They are what Campbell termed "vicarious selectors," which stand in for the real world in the domesticated variation and selection process of learning and cultural change.

The puzzle, then, is how to reconcile the necessary appeal to biological history for the reliability of the senses and the brain with the apparent open-ended creativity of human beings. This is just the same problem that inspired the notion of the selfish meme. In terms of cultural evolution, ideas proliferate due only to how well they satisfy the *immediate* determinates of that propagation. In the absence of constraints on the course of cultural evolution by some reliable surrogate for the real world, there is no reason to think that what evolves in culture has any systematic tendency to be beneficial for human beings. To be sure, there may be cases of "cultural group selection" (Boyd and Richerson 1985) according to which cultures with bad ideas die out, but surely error detection is not always, or even usually, that crude. What the notion of vicarious selectors, error detectors, or "success indicators" as I call them allows is the cultivation of evolutionary processes on the level of individual and cultural learning. This is the key to the puzzle. Because creativity exploits random elements, it is unbounded by genetic history. Because error detection is statistically correlated with the prospects of human well-being, the evolutionary process can gradually and systematically improve learned innovation in the service of human flourishing well beyond the bounds of what was specifically provided for in the history of biological evolution. This is not to say that what individuals learn or culture retains *inevitably* serves human well-being; there are too many obvious examples to the contrary. What needs to be explained is not how culture is *always* good for people but how it manages to be good enough for people, often enough, to overcome the obvious costs involved. The answer is that human nature is such that people have a tendency to stop doing what is bad for them. This tendency takes on a variety of implementations, from simple pain mechanisms, to the imitation of success, to complex assessment of what does and doesn't make sense in the stories we are told.[1]

The principle of the relationship is clear enough and has been for some time. What has been missing is a way to formalize it, to quantify it, to make it approachable via scientific methods. The work we have done so far in this book has not been in vain, however. Equipped with general notions of evolution and information transfer, we can take the final step toward building a proper model for evolutionary epistemology, one which formalizes the tracking relationship between learned or acquired behavioral states and states of the real world which bear on human flourishing.

INFORMATION AND SELECTION ON TWO LEVELS

Formally, one of the ways that the dynamics of two populations can be related is that distributions in one population can effect fitnesses in other populations – for instance, if individuals in one population can form part of the environment for individuals in the other. Competition, coevolution, and predator-prey relations have this feature. In the immediate case, we have a population of traits which are genetically determined, or innate, and a population of traits which are cultural – acquired via learning. The proposed solution to the problem of cultural reliability as stated here is that there are *innate* traits which operate as selection mechanisms on the population of variable *cultural* traits, so as to increase the information about the world in the distribution of the variable traits. These are just Campbell's vicarious selectors, but I refer to them as "success indicators" in what follows.[2] Let's begin with a model of a similar system in a very simple organism.

The suggestion that we can learn something about human knowledge by examining the physiology and behavior of a bacterium may strike some as absurd, but there are two reasons it seems appropriate. One is that it is good to start simply, and the system by which *Escherichia coli* governs its motility can be described in terms very close to those with which we are going to describe the system of controls on human cultural evolution. A great deal is known now about how this system works, which tends to reduce the obscurity and abstractness of the relationship we want to examine. The second is that, since I want to claim the ubiquity of success indicators and their ability to govern phenotypic variability, what better way to demonstrate this than to show similar functioning in the simplest of organisms?

E. coli is a motile bacterium, capable of governing its mobility in such a way as to avoid certain dangers and to locate food (Hazelbauer, Berg, and Matsumura 1993; Manson 1990). Motility is accomplished through a number of rigid, corkscrew flagella, each of which rotates via a rather surprising arrangement involving something that looks a lot like an electric motor and attached by a universal joint. The flagella can be rotated in either direction, in unison, with counterclockwise rotation producing straight swims and clockwise rotation producing a random "tumble." In neutral conditions, the flagella alternate between counterclockwise and clockwise rotation, straight swimming and tumbling, with about equal probability. *E. coli* is also supplied with a variety of chemical sensors, arranged apparently at random on the outer surface of the cell wall. The substances to which the sensors are sensitive can be distinguished as either attractants or repellents, depending on their effect on the bacterium's motility. The sensors are associated with a sort of chemical "memory," which allows the organism to compare current levels of sensor activation with recent levels of activation, so that gradients of attractant and repellent can be detected. When the level of activation of an attractant sensor is increasing, tumbling is suppressed; when the level of activation of a repellent sensor is increasing, the likelihood of tumbling is increased.

Consider a sizable colony of *E. coli* on a gradient of some repellent, a substance to which prolonged exposure is fatal. We can view the colony of bacteria as instantiating two formal populations: a population of genotypes or innate traits and a population of current phenotypic variable traits. So, on one hand, there is the distribution of various configurations of flagella, sensors, and the mechanisms which determine how sensor states affect the direction of rotation of the flagella. On the other hand, there is the distribution of current behavioral states.

We could think of the bacteria of having just two behavioral states, clockwise and counterclockwise, or, if you prefer, tumble and straight, but since the behaviors of interest here are directions relative to the repellent gradient, it would be better to characterize their behavior that way. If we imagine them

Information and Meaning in Evolutionary Processes

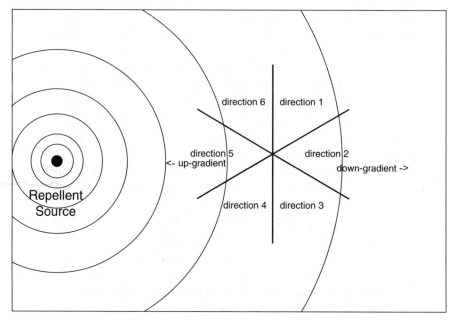

Figure 6.1. Six types of directional behavior.

confined to a plane, we can (somewhat arbitrarily) divide their 360 degrees of options into six segments (Figure 6.1)[3]: straight up-gradient, straight down-gradient, and the four segments adjacent to those two. Which behavioral states get associated with individual genotypes obviously will determine the reproductive prospects of the various genotypes. Which genotypes are present in turn determine the way in which the various behaviors are selected relative to the current environment. It is this interactive relationship that is our central concern here.

Let's think about how the selection on the distribution of variable behavioral traits (directions) works. In neutral situations, where none of the sensors are activated, the bacteria travel in "random walks," alternating between straight swims and random tumbling. In a large population, we can expect that the various directions are equally represented. In such a "flat" distribution, information will be zero (assuming that various environmental states are equiprobable given this distribution). When the colony is placed on a repellent gradient, so that the frequency of tumbling is increased in the bacteria swimming up-gradient and tumbling is suppressed in bacteria swimming down-gradient, the relative frequency of down-gradient swimming will go up. In this case, what happens to the information in the distribution of behavioral traits?

There are two answers to this question, since there are two environments with respect to which information might be increased. One is the environment as determined by the immediate selection pressures on the various directional behaviors, determined by the controls on the relative frequencies of tumbling and straight swimming. In our situation, up-gradient swimming is being locally selected against and down-gradient swimming is being selected for. These fitnesses are determined by the distribution of genotypes in the colony and the resulting configurations of sensors and rotations in combination with the current states of the sensors. By analogy with the human knowledge system, however, this information is simply information about which direction the colony currently prefers, rather than being information about the world. The epistemological question is about how information about the world makes it into the distribution of acquired behavioral traits.

What we need then is a way of characterizing the distribution of behavioral traits in terms of the fitnesses of genotypes, which in this situation is not hard to do. What we can do is say that the distribution of directional behaviors is associated with two fitness quantities which determine two "most fit types." The first is determined by the "local" selection pressures on behaviors, as determined by the distribution of sensors and so forth in combination with the current state of the sensors (i.e., the factors actually expressed in the behavior population's fitness vector $\vec{w}$). We will call this the "most fit_L" behavioral type, with the "L" subscript standing for "Local." The second is determined by which type of behavior would currently make the biggest contribution to the reproductive success of the genotype carried by the individual. We will call this the "most fit_G" type, with the "G" subscript standing for "Genetic fitness contribution." (Recall that we can specify fitness contributions even when there is only one genotype remaining, since our fitnesses are defined as numerical reproductive probabilities.) If the apparatus for selecting behaviors is doing its job, the type of behavior which would make the largest contribution to genetic fitness will be the one that is actually preferred in the local selection process.

So, if the most preferred (fit_L) behavior is always the same as the most fit_G then the conditional probabilities with respect to states of the local environment and states of the genetic environment will match, and the information about each of the two environments will be the same. If, for instance, the most "fit" behavior (in both senses) is direction 2, and if it attains a relative frequency of 0.7, and if this frequency is an invariable indicator that a direction is the most fit (in both senses), then the population will contain 2.58 bits of information about the real world – that is, about the world as it impacts organismic reproduction.[4] Moreover, these gains are reliable.

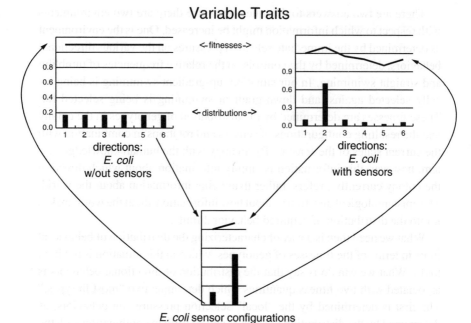

Figure 6.2. *E. coli* selection network.

What accounts for this information gain is the history of selection between genotypes. If, for instance, some genotype determined phenotypes which had no sensors for this particular repellent, or mistakenly interpreted it as an attractant, then it would be displaced by a genotype which did better at correlating adaptive behaviors with the immediate presence of this particular toxic substance (Figure 6.2).

The same reasoning applies to heritable variations in sensitivity to the repellent. Thus, the *accuracy* of matching the most fit$_L$ type and the most fit$_G$ type will increase over the course of genetic evolution where (1) there are variations in accuracy, (2) the benefits of increased accuracy are not outweighed by rising costs, and (3) variation rates are not too high. The result of this increase in accuracy driven by genetic selection should be an average increase in information about the world held in the distribution of variable traits.

The nice thing about the *E. coli* system as an illustration of a basic multi-level selection process is that what happens on the level of the variable traits (which direction they swim) is the generation of currently adaptive behaviors via a variation and selection process. The genetically determined "reflex"

Multilevel Information Transfer

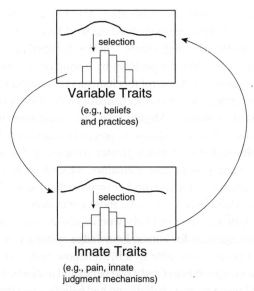

Figure 6.3. Schematic of a basic two-level epistemic control system.

in interaction with the environment results in the selection between various directions but does not determine which behaviors will be available to be selected. These are generated via the random-tumbling process. Figure 6.3 shows a schematic of the basic relationship of this simple two-level system. The distribution of innate traits governs the changes in fitness of variable traits, relative to the current environment.[5] In epistemological terms, over the course of the history of *E. coli*'s genetic evolution the species has "learned" how to determine which directions it is better to swim in a current environment. This knowledge is innate. In particular circumstances, the colony, by application of this innate "knowing how," learns which direction is currently the best one to swim. The random generation of new directions via tumbling is essential to this process, so that this is not simply the expression of an innate trait, but something that looks just like trial-and-error learning.

HUMAN KNOWLEDGE

The basic human epistemological problem, seen through the lens of evolutionary epistemology, looks an awful lot like the problem that *E. coli* has solved. Ordinarily, one might ask, how is it that we can know anything about the world, or how is it that our beliefs and theories are able to come into conformity with the world (assuming they do)? In our terms, the question becomes: how is it that information about the world ends up contained in the distribution of our theories and beliefs – in the distribution of our acquired cultural traits? *E. coli* solves the problem by having certain innate traits, the configuration of sensors and control mechanisms for flagella rotation, which are adapted to function as selection mechanisms on the distribution of variable traits, directions of swimming. Mightn't we have something similar?

The suggestion is that our solution is precisely that which *E. coli* figured out long ago, although one of much greater complexity. Human beings are equipped with a variety of success indicators, which have been designed by evolution to function as selection mechanisms on the distribution of cultural items. Now, this claim may seem to be more controversial than it really is, and I will deal with a number of likely worries about it presently. First, let's consider the consequences for human knowledge systems if it is true.

Suppose that people were able to tell, somehow, with fair reliability the difference between a good belief and a bad belief and instinctively rejected the bad ones. What I mean by good beliefs and bad beliefs is just that good beliefs are ones with genetic fitness consequences that are, on average, positive, and bad beliefs are ones with genetic fitness consequences that are generally negative, given the job that beliefs exist to perform. This pattern of rejection would result in a distribution different from the random one. As a result, exploitable information about the world would end up contained in the distribution of beliefs.

Just what is meant by "information" here, and how much information would end up in a collection of beliefs? With definitions like those explored in Chapter 4, the amount of information contained in the distribution is going to depend on several factors, including the conditional probabilities associated with the beliefs, how many beliefs there are, and how they are "arranged."

Suppose, for simplicity's sake, that there are eight beliefs we can have. If none of these beliefs contradict each other, then each individual can have any, all, or none of them. In this case, we measure the frequency of each belief in comparison to the frequency of not having that belief. It takes one bit to designate one member of a pair, so that for each belief – if a distribution guarantees that one of the pair is the most fit$_G$ (and the environmental states are equiprobable) – then the distribution contains one bit. For our eight beliefs, this means we can have a total of eight bits. If our belief-choice mechanisms are less efficient or reliable, then we may have less information than that. On the other hand, if the states of the environment are not equiprobable, we could end up with a lot more information than that with the same number of belief options.[6]

On the other hand, if all of the beliefs contradict each other, such that an individual can only have one, the situation changes somewhat. Suppose that each member of the community has one and only one of the eight beliefs. If the distribution of beliefs were a sure indication that one among the eight were the most fit$_G$, then the distribution would contain $\log_2 8 = 3$ bits of information – again, assuming equiprobable states. Notice that in this case we have less information for the same number of beliefs. Here, however, each person only has to hold one belief, instead of up to eight, so that information may be held more efficiently on the individual level. Of course, what we have in the real system of human beliefs is large numbers of beliefs, arranged in variously sized sets of contraries.

This is not the place to pursue the analysis of actual human knowledge systems and the way in which they gain and hold information about the world, because the question of actual information gain is largely an empirical matter. The point to be made here is that if evolution has provided us with success indicators, we can use the formal relationship between two populations to model rigorously the way in which their interaction facilitates information gain about the world in the distribution of our beliefs and commitments. This may not give us epistemic justification for particular beliefs or commitments, but does go some way toward justifying the confidence we feel in our overall methodology of inquiry, as well as the overall results of their application.

COMMON SENSE

The success indicators required to form the essential epistemic link by which information about the world ends up in the distribution of our beliefs and commitments may be found, I suggest, in the domain covered by the term "common sense." Now, common sense is an odd subject, appealed to often, but rare is the felt need to explain it or justify our overall confidence in it. The reason I bring it up is not to attempt to rectify this situation but merely to use the term to point our attention toward an aspect of human nature with which we are all familiar and to which all of us, on a day-to-day basis, trust our lives. The success indicators central to this chapter are just the sort of things that we all take common sense to be, or so it seems to me. The most I need to claim here is that people can, with fair reliability and in ordinary situations, *tell* what is good and bad for them – that they are pretty good at recognizing the most fit$_G$ from among a group of alternative behaviors. We are good at making certain kinds of judgments, and this ability is part of our genetic heritage, fully as much as *E. coli*'s hardwired responses to chemical gradients is part of its genetic heritage. This is not to say that all of common sense is innate, since much of what comes under the heading "common sense" is clearly learned behavior. (Nor is this to ignore the fact that environmental changes, such as the increased availability of salt and fat or the availability of pleasure-inducing drugs, compromise the reliability of the benefit indicated by success indicators.) It seems equally clear, however, that there is some subset of common sense that is truly innate and that retains much of its reliability.

Here, we're interested in the innate variety. Consider, at one end of a scale of complexity, the simple fact of pain caused by physical trauma. I believe that stoves are safe to touch. I touch the stove. I feel pain. The belief is inhibited. Beliefs with behavioral consequences that become associated with pain get inhibited – they are selected$_L$ against. And it's a good thing, both for me and my future offspring. What's going on here is similar to what goes on in *E. coli*. To anthropomorphize a bit, what counts as pain for *E. coli* is whatever process is initiated by rising activation of repellent sensors; what counts as pleasure is whatever process is initiated by rising activation of attractant sensors. All vertebrates have nervous systems sophisticated enough to have recognizable "pain" reactions, and in each case the purpose of these mechanisms is one that we should now recognize as essentially epistemic. Within our framework, the function of pain is to detect "error" in trial-and-error processes. It is to facilitate increases in information about the world in distributions of variable traits.

Not all human success indicators have the sort of success-failure indicator quality that human pain shares with the detection mechanisms of *E. coli*. Quine, in "Natural Kinds" (1969), argued that our *recognition of similarity* must be innate if we are to learn anything at all and that evolution has adapted our cognitive similarity space to similarity relations in the world. Presumably, *memory* serves the same function – we need to be able to compare current states to previous states, just as *E. coli* needs its own primitive "memory" to detect gradients. *Imitation* is governed by our ability, call it "empathy," to recognize whether a behavior has caused pain *before* we try it. Hume suggested that our tendencies to make inductive and causal inferences are tendencies nature has provided. Hume, however, could not explain how this was, writing a hundred years too early for any help from Darwin. Understanding something about evolution, we are now in a position, with Waddington, to explain how Hume's version of "preestablished harmony" actually gets established.

> The faculties by which we arrive at a worldview have been selected so as to be, at least, efficient in dealing with other existents. They may, in Kantian terms, not give us direct contact with the thing-in-itself, but they have been moulded by

things-in-themselves so as to be competent in coping with them. (Waddington 1975, 36)

Once again, at this juncture I only mean to gesture toward a region of human nature with which we are all quite familiar. It seems to me rather obvious that humans have success indicators – lots of them – and most people seem to share this intuition. There is a tradition, which may have reached its height with G. E. Moore's (1962) proof of the existence of his hands in "Proof of an External World," which seems to share our conviction that the solution to the problems of epistemology lie in common sense.

OBJECTIONS

I promised that I would address some worries about this approach to epistemology and the matter of innate human behaviors. Now is the time.

1. Much of the adverse reaction to E. O. Wilson's *Sociobiology: The New Synthesis* (1975) and *On Human Nature* (1979) and to the sociobiology program in general stems from implications regarding limits on human nature. The implication of the fact that some human behaviors are innate is taken to be that there are some features of human nature that cannot be overcome by social conditioning, education, and so forth. What disturbs many people about this in the current political climate is that it makes the prospects of true racial and gender equality seem rather dim.[7] Since then innate human behaviors have been something of a touchy subject, one that the prudent would do best to avoid. Whatever the actual implications of sociobiology vis-à-vis these worries, the kind of innate behaviors required for the solution to the problem of epistemic access offered here does not involve limits on human behavior, but rather powers. To put the matter in familiar language, common sense may be, to some extent, part of our genetic heritage, but this does not mean that we have no choice but to follow its dictates or that differences in the natural endowment of common sense lead with any necessity to differences in capacities on the individual level.

2. I have been using the term *innate* in preference to the more cumbersome *genetically determined*, but both terms are subject to the same criticism: that human behavior is characterized by its flexibility, and there may be no behaviors that are shared by all human beings who carry some particular gene. If a behavior is not shared by all carriers of a genotype, how can it be "innate"? The answer here is that what is required for the purposes

of epistemology is not that a particular success indicator is shared by all gene carriers, that every such individual has some such mechanisms, or that when they are present they are completely reliable. All we require is that, due to the distribution of genotypes in the community, the distribution of variable traits evolves so as to increase the information about the world it holds because of selection on acquired traits by the success indicators. This is consistent with partial reliability of success indicators, uneven distribution of the mechanisms among individuals, and the presence of individuals who are incapable of making any contribution to the information gains in the distribution of beliefs at all. What "innate" means in this context is that under certain historically "normal" conditions, most individuals develop the success indicators in question because of the unfolding of the genetic potential in the presence of those conditions. The fact that environmental abnormalities may compromise the development or expression of those abilities does not cut against the fact that these abilities are what the genes are supposed to result in.

3. It might be objected that if I have managed to offer a solution to the "problem of epistemic access," it is only because I have given that name to a different and simpler problem than it is usually taken to be. I have not, for instance, said anything about how information of the world gets captured in the syntactic structure of beliefs and theories, and this is surely what we meant when we asked how information about the world gets into our beliefs and theoretical commitments. This seems to me to be a fair criticism, as far as it goes. My formulation of the problem of epistemic access – "how does information about the world end up in the distribution of human variable traits?" – is a somewhat different question than the traditional one, and it is an easier question as well. It has the advantage of being soluble. I'm not prepared to argue the point at this time, but the position I take is that information about the world is not supposed to be captured in the syntactic structure of beliefs and theories but in their distributions. The role of syntactic structure is complex and varied, but it is not to hold information about the world. Syntactic details function to structure the space of phenotypic variability, and to control and regularize the behavioral consequences of particular beliefs and theoretical commitments. The case is similar to that of information in the gene pool. Information about the world is not contained in the "syntactic structure" of individual genes or genotypes, but in the distribution of genes subsequent to selection. The function of syntactic structure is to code for particular proteins, which have particular consequences for

development, morphology, and behavior. The syntactic structure of genes also determines which genes can mutate into which other genes (structuring the space of genotypic variability). This general view seems to me to be a natural consequence of using evolutionary theory to do semantics (Cf. Millikan 1984).

4. In the account so far I have not been using the term *knowledge*, but the implication is that propositional knowledge will consist of commitments to the most fit$_G$ from among a set of contrary beliefs. This suggests an identification of best genetic fitness consequences with truth, but this can't be right, since we all know that in particular circumstances false beliefs can be more conducive to survival and reproduction than their true alternatives. The simple answer to this problem is that which belief is the most fit$_G$ is assessed on a probabilistic basis rather than on an individual basis. Thus, it is also the case that in particular circumstances, some belief other than the most fit$_G$ may contribute the most to an individual's reproductive success. But does this mean that I *am* identifying truth with the most fit$_G$ from among a set of contrary alternatives? This is where the issue gets more complicated. My preferred account of "truth" is discussed in Chapter 8.

5. Finally, while success indicators may explain how information about the world gets into the distribution of simple beliefs such as "you'll get burned if you touch a hot stove," it is less plausible to suggest that they explain our confidence in the choices we make between alternative scientific theories. The reliability of the success indicators regarding simple everyday beliefs derives from the fact that the history of natural selection on the judgment mechanisms involved just those sorts of beliefs. On the other hand, we cannot have been provided by nature with primitive mechanisms for adjudicating between scientific theories of (say) subatomic physics, since such theories were not part of the ancestral conditions in which the reliability of those mechanisms was tested. More briefly, we cannot be provided with judgment mechanisms which are reliable in choosing between scientific theories in the way that common sense is reliable in dealing with everyday matters.

This is obviously right in that the solution to this problem involves a relationship that has not been addressed here. It is, however, well within the capability of the multilevel selection model. The basic idea is that predictive consistency with basic success indicators forms part of the selection regime for theories and abstract claims. Similar considerations hold for abstract entities (e.g., electrons). Articulation of such a model is the next logical step.

7

Information in Internal States

The final, critical step in developing the information-and-evolution framework is to understand how selection drives informational tracking of world-states by the *internal* states of neurological control systems such as human learning. The bacterial navigation model allowed us to understand the essential fitness feedback loop – the way it results in acquired traits tracking environmental changes and the way it effectively directs open-ended creativity toward the solution of evolutionary problems. The shortcoming of that model was that the acquired traits involved were implausible analogs for human beliefs, in that nothing the least bit like memory was involved. If the upper level is interpreted as consisting of beliefs, then the all-important interaction with the world via resulting behavior was left out of the model. In this chapter, we add a third, intermediate level for *preference formation* to the model, which will bias variation in a random walk behavior pattern. We again model a system much simpler than humans – bumblebees – but the important thing here is the principle of information transfer involved. Again, the less we assume about complex structures, the better off we are in terms of the epistemological challenge. If information transfer depends on certain structures, then more complex systems which have those structures as well should be able to exploit information in the same manner.

THE MODEL: REAL'S BUMBLEBEES

For a number of years, behavioral ecologist Leslie A. Real (1991, 1992) has been studying the foraging behavior of bumblebees in enclosed environments. In Real's experiments, the bees are confronted with a variety of colored plastic flowers, each of which may or may not contain a precisely measured quantity of nectar. For each distribution of nectar that is tried, the rates at which the bees visit flowers of each color are recorded and analyzed. Real is concerned with issues such as risk aversion in the bees' foraging, so that common manipulations involve flowers of different colors having the same mean amount of nectar, but different variances. The idea is to manipulate these quantities to determine how important consistency in payoffs is to the bees – how much of a trade-off in terms of average nectar levels they are willing to accept for greater regularity in consumption. Some work has also been done on the range of sensory cues that *honey* bees can respond to in learning, which bears some relevance to the argument developed herein. In particular, many of these experiments use simple rectangles of colored plastic as "flowers," rather than anything we might suspect bears any close resemblance to flowers from the bees' ancestral environment (see Lamb and Wells 1995; Petrikin and Wells 1995).

The bumblebee-foraging model has been productive in facilitating focused questions about both optimization and the mechanisms of learning, because of its combination of simplicity and sophistication. For instance, Montague et al. (1995) have explicitly modeled (via computer simulation) a mechanism that facilitates the kind of learning that Real has been investigating empirically. The proposal involves a simple neurological reinforcement mechanism that depends on the global release of a neurotransmitter, which modifies the "weights" of the associations between colors and foraging tendencies. Neural connections between color sensors and neural sensors result in navigational patterns that increase the probability of foraging at flower colors most associated with activation of nectar sensors, resulting in the adaptation of current learned behaviors to prevailing conditions. It is interesting to note that the neurotransmitter involved is closely related to dopamine, which apparently plays a similar role in learning in higher animals, including humans, because this may indicate that on the neurological level some aspects of human learning are not that different from bee learning (see Montague, Dayan, and Sejnowski 1996).

In what follows, we shall be asking a slightly different question than those of either Real or Montague et al. – to start with, how efficiency in tracking a changing environment results from selection on genes and, then, more important for our point here, what difference novelty in environmental states and internal representations makes for this efficiency. Note that we shall be explicitly modeling this process as a multilevel selection process in Campbell's (1974) sense, in contrast to the neurologically based model of Montague et al.

Let us begin with a two-level model for "simple" bumblebees which will reiterate more formally the structure discussed in Chapter 6. Suppose that our scientist manipulates the bees' environment in the following manner: there are two kinds of flowers, blue and yellow, and there are equal numbers of each kind in the enclosure, spaced evenly and distributed at random. At any given time, either all the blue flowers are full and the yellow empty, or the yellow are full and the blue empty. Moreover, our scientist changes which kind of flower is full randomly, but at some specified rate. So, for each discrete-time cycle, there is a certain probability (the "environmental fluctuation rate" **dE**) that the environment will change which state it is in, blue (S_B) or yellow (S_Y). Let us also assume that going to a full flower takes twice as long as going to an empty one, so that foraging at a full flower takes two cycles, and foraging at an empty flower takes only one cycle. (For simplicity's sake, we will assume that the flowers are refilled immediately after the bee leaves, so we can neglect the depletion of nectar over the course of a day.)[1]

Here is the foraging strategy of our simple bees. They go to a flower at random, feed if it is full, and then go to another chosen at random. Our simple bees are incapable of learning which kind of flower is better to go to on a given day, they are incapable of even acquiring a preference for one over the other without having the preference arise and be selected via the evolution of their genes. So there are two kinds of foraging behavior, which we will designate as β_B for foraging at a blue flower and β_Y for foraging at a yellow flower. The change in the relative frequencies of these two behaviors will constitute the evolution of the upper level behavior distribution, and the change in relative frequency of the simple bees' genotype **g** compared with its competitors (which we need not designate here) will constitute evolution on the lower (genetic) level. To keep the subscripts manageable, we will designate these respective frequencies as $p(\beta_B)$, $p(\beta_Y)$, and $p(\mathbf{g})$.

Figure 7.1 provides a useful heuristic for this system. It depicts a two-level "adaptive landscape," in which the bar graph indicates for each level the relative frequencies of different types included in the distribution, and the line graph above indicates the relative fitnesses. The upper-level distribution of "foraging behaviors" represents the distribution of behaviors carried by genotype **g** at a given moment. A fuller representation would show behavior distributions for each of the competing genotypes. Long arrows indicate important functional dependencies. In the case of Figure 7.1, relative genetic fitness is dependent on the distribution of behavioral types – how well a genotype does depends on whether the bees carrying the genotype are going to full or empty flowers. Fitness, again, is just the propensity of a type to increase or decrease in the current environment, due to casual interactions of tokens of the type in question. We shall designate these fitnesses $w(\beta_B)$ and $w(\beta_Y)$, and we should recall that these are "local" fitnesses in order to emphasize that they are not measures of the probable genetic fitness contributions of these acquired

Information in Internal States

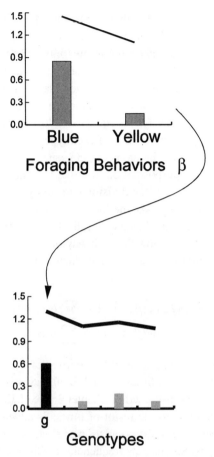

Figure 7.1. Two-level selection network.

or transient traits but simply represent proximal factors driving the evolution of the belief population. (Notice that heuristics such as Figure 7.1 depict these local fitnesses rather than the probable genetic fitness consequences of behaviors β.)

Informally, the dynamics of the behavior distribution are as follows: (local) selection on behaviors is due to (1) the bees' innate sensors which tell them to go elsewhere when there is no (more) nectar in the flower, and (2) the mechanical constraints on the time it takes to determine whether a flower has nectar. Here, it is tempting to think of all selection as resulting in differential "mortality" in the sense of the differential cessation of behaviors as opposed to differential reproduction or dissemination of the behavior. (We suppose that the bees don't imitate or even follow each other.) However, since the bees are

Information and Meaning in Evolutionary Processes

always going somewhere, the distribution has a fixed size, within a biological generation. Thus, it is probably more perspicuous to characterize the process as one of mutation, in which each bee has one foraging behavior token, which changes type over time.[2] Thus, we have a distribution in which the dynamics are entirely those of selection-driven mutation (see equation [9] in Chapter 3). The biologically counterintuitive constant size[3] of the distribution means that the mutation rate for each type is equal to the loss to selection, with equal probability of generating a new token of each type.

We assume that the bees' innate nectar-presence sensors are 100 percent accurate and that for the purpose of our discrete-time model, the bees' visits are synchronized. This means that if visiting an empty flower takes one cycle and visiting a full one takes two, the local fitness $w(\beta_B)$ of the blue flower behavior is 0.5 when the blue flowers are full (S_B) and 0.0 when they are not (S_Y), and likewise for the yellow flower behavior β_Y.

Given these assumptions, the new frequencies $p(\beta_i)'$ of the foraging behaviors are

$$p(\beta_B)' = p(\beta_B)w(\beta_B) + (1 - \sum p(\beta_i)w(\beta_i))/2$$
$$p(\beta_Y)' = p(\beta_Y)w(\beta_Y) + (1 - \sum p(\beta_i)w(\beta_i))/2,$$

with $i \in \{B,Y\}$. The first term is the old proportion of the particular behavior times that behavior's current fitness, which in all cases is less than 1 (since there is no imitation). As discussed in Chapter 3 (see "Selection-Mutation in Fixed-Size Populations"), to the result of fitness is added the second term, which is the total decrease in the population as a whole, divided by 2, which is the number of types. So, the new frequencies are the old frequencies times their fitness (in this case, the local "destabilization rate" of the individual behavior tokens) plus their share (half) of the distribution's loss to selection in the current cycle. This maintains the population size, with uniform mutation into the two types, and is an accurate model of the "random walk" dynamics of the bees' behavior distribution.

At this early stage in model development, we can predict its behavior analytically. While the environment remains the same (i.e., the distribution of nectar among flowers remains the same), the distribution quickly approaches an equilibrium state where two-thirds of the bees are foraging at full flowers and one-third are foraging at empty flowers. This inefficiency, which holds even on the assumption that the bees' nectar-presence sensors are perfectly accurate, will of course constrain their ability to compete. If the impact of foraging efficiency on genetic fitness is proportional to the relative amount of time they spend at full and empty flowers, then our simple bees' genetic fitness in S_i will be

$$w(\mathbf{g}) = p(\beta_i) + \mathbf{C},$$

where $\mathbf{C}$ is the constant background fitness that the simple bees share with their competitors. So, the contribution of foraging behavior to genetic fitness is only two thirds of what it could be if the bees were foraging more efficiently. (Note here that genetic fitness $w(\mathbf{g})$ is not an absolute growth rate but a relative fitness.)

PREFERENCE FORMATION: ADDING A THIRD LEVEL

The moral so far is that the random-search strategy of the simple bees places severe limits on their foraging efficiency because of the relative amount of time it takes to determine that a flower is empty.[4] Any bee-variant which can improve on the efficiency of the random-search strategy (without incurring offsetting costs) will have higher genetic fitness and thus eventually displace the simple bees. Moreover, because of the repeated visits to flowers of different colors and the bees' accurate perception of the presence of nectar in those flowers, it seems that the "information" needed to improve on this strategy is available. We need a third level, however, to represent the process of learning which flowers currently hold nectar.

Suppose that a new bee variant arises which can form "preferences" (π) for flower color and that the strength of those preferences affect to which color of flower they go after they have finished with the last one. Since there are two colors, there will be two preferences, π_B and π_Y. To be adaptive, the strength of those preferences needs to vary over time with the recent success of individual bees in finding nectar. This introduces a third (middle) level into our selection hierarchy (see Figure 7.2).

In this case, the particular selection mechanism on preferences π will be borrowed from those on behaviors β – the functioning of the innate nectar-presence sensors in combination with the actual presence of nectar, although relative foraging time is not a factor; that is, the fitness of the preferences $w(\pi)$ is not sensitive to the amount of time it takes to forage at full versus empty flowers in the way that the fitness of the behaviors $w(\beta)$ is. It may be helpful to think of this as the bees "associating" flower colors with nectar, where the strength of the association is a function of the recent co-occurrence of color recognition and activation of the nectar-presence sensors. The result of the process is a biased random variation in the evolution of the distribution of foraging behaviors β. This constitutes a departure from Campbell's

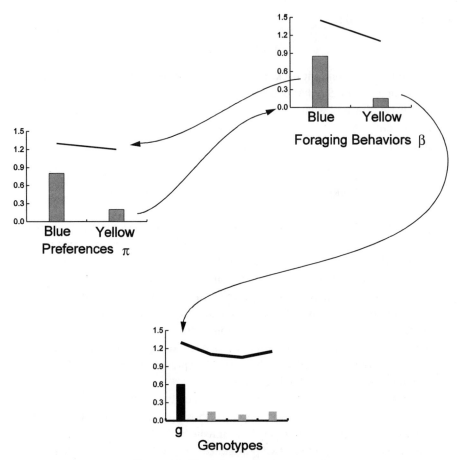

Figure 7.2. Three-level selection network.

(1974) scheme, more closely resembling the "guided variation" described by Boyd and Richerson (1985).

We need not concern ourselves with exactly how these preferences are constituted and updated on the individual level; what is important is the overall effect on the relative frequency of behaviors of the smart variant.[5] Consequently, by the frequencies $p(\pi_i)$, we shall designate the overall tendency[6] of the genotype currently to visit a blue or yellow flower when in the initial phase of the foraging cycle, and by the fitnesses $w(\pi_i)$ we shall designate the expected rate at which these tendencies increase or decrease because of the effect of their own presence (i.e., their effect on the distribution of the β_is). Characterizing preference frequencies $p(\pi_i)$ in this way allows us to simply substitute the frequency into the mutation term of the equation given previously for the behavior dynamics, so that

$$p(\beta_i)' = p(\beta_i)w(\beta_i) + p(\pi_i)(1 - \sum p(\beta_j)w(\beta_j)).$$

Instead of giving $p(\beta_i)$ half of the distribution's loss to selection, we give it $p(\pi_i)$ of that loss.

The equations governing the dynamics of the π distribution will necessarily look a bit different from those that govern the β dynamics. This is because there is no reason to suppose that the reinforcement patterns on preferences π are such that the total strength of the two preferences combined always sums to some constant. Consequently, the evolution of the π distribution needs to be governed by the more familiar "replicator dynamics," appropriate to populations of varying size.

Recall that the frequency-independent replicator dynamics in discrete time have the form

$$p_i' = p_i w_i / \sum_j p_j w_j,$$

so that for any type i, its new frequency is its old frequency times its fitness divided by the current population mean fitness. If the fitnesses are absolute growth rates, then mean fitness is just the total increase in the population size in the last generation, and dividing by it "renormalizes" the frequencies of the various types so that they sum to one. Adding mutation to the equation involves adding the gains from mutation and subtracting the losses to mutation from the result of the basic replicator dynamics.[7] (The reason we need to have mutation in the π distribution is that, without it, the dynamics may go to fixation. That is, without mutation, it is possible under the replicator dynamics for one or the other of the preferences to go extinct, so that neither individual bees nor the whole bee population can continue to learn to prefer flowers of the associated type.)

So, adding a uniform mutation rate μ to the distribution of preferences π (think of this as curiosity or doubt), the dynamics of the distribution of π's will governed by this:

$$p(\pi_i)' = [p(\pi_i)w(\pi_i)/\sum p(\pi_j)w(\pi_j)] \cdot (1 - \mu) + \mu/2,$$

where $w(\pi_i) = 1/\mathbf{I} + p(\beta_i)$ if $\mathbf{S}_i$ obtains, and $1/\mathbf{I} - p(\beta_i)$ otherwise, with $\mathbf{I}$ being the "intensity" of the selection on the preferences π. Notice that when $\mathbf{I} = 1$, then $w(\pi_i) = 1 \pm - p(\beta_i)$, depending on which state $\mathbf{S}$ obtains, and the involvement of $p(\beta_i)$ makes the fitness of the preference reflect the fact that it is the performance of the associated behaviors that provides the occasions for testing the appropriateness of the preferences themselves. The higher the value of $\mathbf{I}$, the more the frequency of the associated behavior β_i matters to the preference fitness $w(\pi_i)$, so that **I** is a sort of measure of conservativeness in learning or rate of learning, similar to those used in standard learning models.

Supposing that the fitnesses of the behaviors $w(\beta_i)$ depend on the states $\mathbf{S}_i$ as with the simple bees and that while in state $\mathbf{S}_i$ and for values of $\mu = .01$ and $\mathbf{I} = .05$, the joint dynamics converges to an equilibrium where

$$\hat{p}(\pi_i) \approx .905;$$

$$\hat{p}(\beta_i) \approx .95; \text{ and consequently}$$

$$\hat{w}(\mathbf{g}) \approx .95 + \mathbf{C},$$

which is significantly higher than the corresponding performance of the simple bees (at 2/3 + C). Note that extreme values of **I** and μ may result in only marginal improvement over the performance of the simple bees. On the other hand, lowering the value of μ relative to **I** brings the value of $\hat{p}(\pi_i)$ arbitrarily close to 1, with corresponding gains in the equilibrium genetic fitness $\hat{w}(\mathbf{g})$.[8]

VARIABLE ENVIRONMENTS

Static conditions of the kind that we have been considering so far hardly constitute an appropriate challenge for the learning of our bees since adapting to this sort of static environment is something that one would expect to be handled by genetic mechanisms, rather than the kind of adaptively plastic behavioral changes discussed here. Things become more interesting when the environment varies between $\mathbf{S}_B$ and $\mathbf{S}_Y$, although the mathematical analysis of the resulting system becomes difficult. It is easy, however, to write a computer simulation that models the joint dynamics of the two types of bee systems under varying environments, and compares the resulting genetic fitness $w(\mathbf{g})$ of the smart bees with that of the simple bees. What one finds, and this not surprising, is that under a wide range of plausible values for intensity **I**, mutation rate μ, and environmental fluctuation rate **dE**, the smart bees do better than the simple bees. The reason that this is not surprising is that *any* improvement on the simple bees' random-search strategy – any mechanism that introduces even a little correlation between where the bees go and where the nectar is – will prove advantageous. The only situation in which smart bees do worse is where the preferences reduce correlation with the environment. This can happen with very low values of μ or **I**. (Note that if high costs were involved in the preference-forming mechanisms, this could also offset advantages of preference formation. I omit inclusion of costs from the calculation for simplicity's sake and since assigning values for them would be arbitrary. It is enough to recognize that the smart bees might need to do substantially better than the simple bees to compete, depending on the costs involved.)

The real point of introducing the third level here is not to argue that more complex environmental sensitivity and behavior regulation can be adaptive, but to model the way that information is caught in mid-level regulatory distributions like the preferences of our smart bees, and the way that the exaptation of success indicators from simple systems allows fuller exploitation of the information they generate. The notion of information here, as discussed in Chapter 4, is that of "mutual" information, which measures statistical correlation between states of the environment (S_i) and states of the distribution. Since there are only two types in each of the distributions in the current model, complete descriptions of the states of the distributions can be given in terms of the frequency of only one of the types from each distribution, for example, $p(\pi_B)$ and $p(\beta_B)$. Moreover, to make things manageable computationally, I divided the ranges of these frequencies into ten subranges, which counted as the states of the distributions. So, for instance, if $p(\pi_B)$ was between 0 and 0.1, distribution π was in the first state; if between 0.1 and 0.2, in the second; and so on – similarly, for the behavior distribution β. The environment had only two states, S_B and S_Y, as described earlier.

Something one discovers when constructing models of this sort is that as the complexity of the model increases, the number of parameters which require arbitrary values increases. For instance, the precise equilibrium value given for the simple bee model depended on the relative amount of time it took to forage at the different flowers, and the smart bee model depended in addition on the assignment of values for intensity of preference selection (**I**), and the "curiosity factor" (μ). One might justifiably ask why *those* values were chosen, and the answer must admit that some degree of arbitrariness was involved. Since it is reasonable to suppose that there are a variety of rates **dE** at which the environment varies, and a variety of rates of preference mutation μ and intensity of selection on preferences **I** that might arise as adaptive responses, assessment of the performance of the system in tracking a fluctuating environment must include some exploration of variations in performance across different combinations of values for these three parameters. Moreover, this allows us to assess the covariance of various kinds of information with genetic fitness over a variety of conditions.

Consequently, the simulation was run with a variety of settings of **I**, μ, and **dE**, simulating different values for intensity of selection on preferences (including the accuracy of the associative mechanisms), mutation (or "curiosity") of preferences, and rates of environmental fluctuation. As before, I assumed that the nectar presence sensors were 100 percent accurate, so that the fitnesses of the behaviors were the same for all trials at $w(\beta_i) = 0.5$ if S_i, 0 otherwise. The individual trials were run for something more than 1 million cycles each for the probabilities and, thus, the mutual information measurements to stabilize. Values for population and individual information between both the behavior distribution β ("β-info") and the preference distribution π ("π-info") and the environment, as well as the average contribution of the behavior distributions to genetic fitness $w(\mathbf{g})$ ("g-fitness") were recorded and compared. It was convenient to graph both differential fitness ($w(\mathbf{g}) - \mathbf{C} = p(\beta_i)$, for S_i) and information together since both quantities were confined to the zero-one interval and the maximum average mutual information was 1 bit and the maximum value of $p(\beta_i)$ was 1. The results were as follows.

The performance of the simple bees was used as a basis of comparison. In fact, simple bees were run as a competing genotype in the simulation. Across the entire range of trials, the fitness of the simple bees was right around the equilibrium value of $2/3 + \mathbf{C}$ calculated earlier because the simple bees were at a flower with nectar two thirds of the time. High rates of environmental fluctuation reduced this value somewhat. In the extreme case, variation every cycle ($\mathbf{dE} = 1$) reduced genetic fitness of the simple bees to $0.6 + \mathbf{C}$. In all cases, the information about the state of the environment in the distribution of foraging behaviors β (the "β-info") was quite high, between 0.99 and 1 bit of average mutual information. (Recall that 1 bit is the amount of information generated by the environment, and, thus, the theoretic maximum for mutual information.) Information on the individual level was quite low, however, about 0.082 bits. Thus, the tragedy or, if you like, the potential opportunity, is that the population as a whole "knows" exactly where the nectar is at any given time, but this information is not fully exploitable, resulting in both low fitness and low individual information.

The reason for this combination of high population information and relatively low fitness is that the extreme mutational characteristics of random walk behavior causes the distribution to approach its equilibrium value within about twenty cycles of each change in the environment. This results in the distribution being in the same partition ($.6 < p(\beta_i) < .7$, for S_i) the vast majority of the time that S_i obtained. This increases the probability of S_i given the distribution, and thus the average information. The low fitness is, in this case, related to the high population information because the ability of the behavior dynamics to approach its equilibria so quickly is largely due to the high effective mutation rate, which results in the close proximity of the equilibria to each other in the interior of the space ($p(\beta_B) = 1/3$ and $p(\beta_B) = 2/3$). The very proximity of these values, however, keeps the dynamics away from the regions where higher genetic fitness contributions can be made. The lesson here is that high information can result from *efficient equilibrium-seeking properties* of the β dynamics (e.g., high mutation rates) that may not contribute to genetic fitness. It will be useful to keep the dynamical behavior of the simple bees' behavior distribution in mind, since the smart bees' dynamics tend to converge toward them in situations where their control system breaks down.

It turns out that the addition of the third level, wherein the smart bees borrow the foraging-success indicator from the simple bee system and use it as a selection mechanism on preference strengths, allows fuller exploitation of the information generated (at the population level) in the simple bee system. It is as though the smart bees model the dynamics of their own population in their preferences and use the resulting information as a guide to their own foraging behavior. As calculated earlier, this increases the foraging efficiency of the smart bees. Let's see how it works over a range of settings.

SIMULATION RESULTS

The six graphs that follow demonstrate the performance of the smart bee model over ranges of values for **dE** (environmental fluctuation rate), **I** (intensity of selection on preferences), and μ (preference mutation rate). For each variable, five quantities are assessed. In addition to genetic fitness which appears as the solid line in all graphs, both individual- and population-level information about the external state is assessed for both the behavior population (which is present in the dumb bee and bacterial models) and the preference population (which is the newly introduced third or intermediate level). If we visualize **dE**, μ, and **I** as defining a three-dimensional parameter space, each pair of graphs plots the values of five quantities (genetic fitness plus four informational measures) taken along a line through that space. The three lines through parameter space lines defining the three pairs of graphs intersect at the default values $\mathbf{dE} = .01$; $\mu = .01$; $\mathbf{I} = .05$, which is to say that the six graphs are only a partial indication of the relationship between these parameters and measures, even for this simplest of models.

Consider first how varying the rates of environmental fluctuation **dE** affects the genetic fitness of the smart bees. Figures 7.3 and 7.4 show the results of a number of trials for the smart bee simulation where the rate of environmental fluctuation **dE** was varied and where **I** and μ were held constant at the values used earlier. The first item to notice is that, as one would expect, lowering the rate of environmental fluctuation does, in fact, result

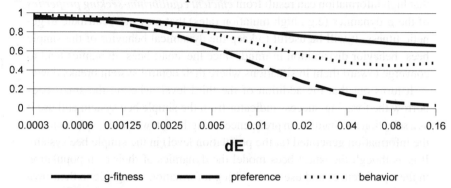

Figure 7.3. Genetic fitness and population information for both preference and behavior populations plotted over rates of environmental fluctuation **dE**. Default values: $\mu = .01$; $\mathbf{I} = .05$. The increase in behavior information beyond $\mathbf{dE} > 0.08$ indicates the breakdown of control of behavior by preferences when environment changes too fast for preferences to track.

in both more information and in more genetic fitness. Notice also that in all of the trials in these sets, the behavior distribution had more information than the preference distribution. This is because the value of **I** is low relative to the effective intensity of selection on the behavior distribution. As a result, the behavior distribution was more responsive to environmental variation than the preference distribution and, consequently, held more information about it, at both the individual and population level. Notice that at the

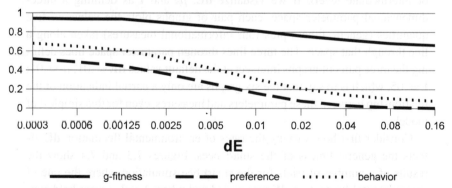

Figure 7.4. Genetic fitness and individual information for both preference and behavior populations plotted over rates of environmental fluctuation **dE**. Default values: $\mu = .01$; $\mathbf{I} = .05$.

population level, preference-information has a closer relationship to genetic fitness than behavior-information; as **dE** decreases, both genetic fitness and population/preference-information increase continuously. This relationship between population/preference- information and genetic fitness is not universal, as we will see, but it does hold while mutation rates μ for the π distribution are fixed. Nevertheless, covariance between fitness and population/behavior-information fails even over changes in **dE**, as evidenced by the minimum in the population/behavior-information curve at **dE** = .08. What happens is that beyond a certain threshold, the regulation of the β distribution by the π distribution breaks down because of the inability of the π distribution to track the rapidly fluctuating environment. As this happens, the dynamics of the β distribution increasingly resemble those of the simple bees. The behaviors of the β distributions of the simple and smart bees converge when the environment varies at each cycle, population/behavior-information reaches a perfect value of 1, and genetic fitness drops to 0.6 + **C**. The minimum in the population/behavior-information curve occurs at lower values of **dE** for lower values of **I** than those used in this set of trials. Notice also that individual information in both preferences and behaviors covaries with genetic fitness in these trials.

One way that an evolving population might try to adapt to a particular rate of environmental fluctuation is by varying the mutation rate μ on preferences, so as to optimize the tracking characteristics of the dynamics. Figures 7.5 and 7.6 demonstrate what happens when mutation rates μ for the π distribution are varied while **I** and **dE** are held constant. The salient feature here is the peak in the value of genetic fitness at $\mu = .02$, which optimizes genetic fitness for the particular rates of **I** and **dE**. The reason for this is that for increasing high levels of μ, the β-dynamics of the smart bees again increasingly resembles those of the simple bees, this time because of relocation of the β equilibrium toward the center of the space. At the upper range of values of μ for these trials, β-info approaches 1 and fitness approaches 2/3 + **C**, the same values as for the simple bees. At the other end, extreme low rates of μ can result in the π distribution going to fixation, with severe negative consequences for genetic fitness. (Average genetic fitness is about 0.5 + **C** when this happens; π-info is, of course, 0.) This is the only case when the "cost-free" smart bees failed to do as well as the simple bees. Notice again that both individual/preference-information and individual/behavior-information covary with fitness.

Recall that lowering the intensity of selection **I** would simulate lowering the accuracy of the selection mechanisms on preferences π, as well as reflect a sort of conservatism in preferences in the face of new information. Consequently,

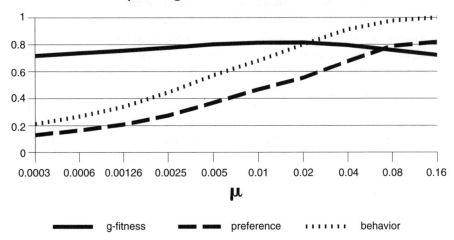

Figure 7.5. Genetic fitness and population information for both preference and behavior populations plotted over rates of preference mutation μ. Default values: $\mathbf{dE} = .01$; $\mathbf{I} = .05$. Optimal prference mutation rate ("curiosity levels") is indicated by genetic fitness optimum at $\mu = .02$.

various values of **I** constitute a different sort of optimizing response than the differing values of μ considered earlier – more straightforward, one might think, but also more difficult to achieve, since high values of **I** require high accuracy in the response mechanisms that select the preferences. Figures 7.7

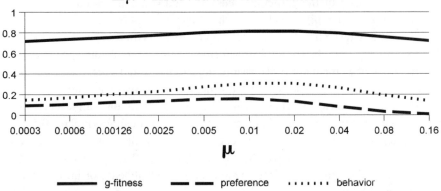

Figure 7.6. Genetic fitness and individual information for both preference and behavior populations plotted over rates of preference mutation μ. Default values: $\mathbf{dE} = .01$; $\mathbf{I} = .05$. Notice that individual information covaries with genetic fitness, unlike population information.

ΔI : Population information

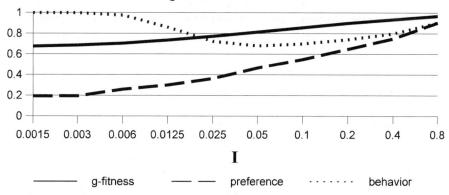

Figure 7.7. Genetic fitness and population information for both preference and behavior populations plotted over levels of selection intensity on preferences **I**. Default values: **dE** = .01; μ = .05. The combination of high behavior information and low individual information for low values of **I** indicate the population acting essentially as simple bees without guidance from preferences. Preference information covaries with genetic fitness.

and 7.8 demonstrate what happens when the intensity of selection on the π distribution (including accuracy) is varied while μ and **dE** are held constant.

The notable feature in Figure 7.7 is the dramatic minimum in population/behavior-information at **I** = .05. This is because of the progressive domination of the π dynamics by mutation as intensity of selection decreases. At very low values of **I**, this results in preference frequencies $p(\pi_i)$ hovering

ΔI : Individual information

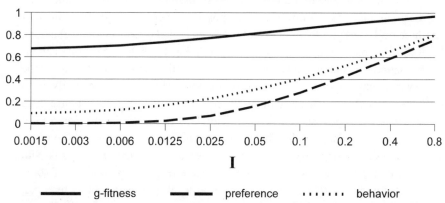

Figure 7.8. Genetic fitness and population information for both preference and behavior populations plotted over levels of selection intensity **I**. Default values: **dE** = .01; μ = .05.

around .5, just as when μ is high, so that the β dynamics of the smart bees again resemble those of the simple bees because of their similar mutational characteristics. The location of the minimum in β-info is sensitive to values of μ, occurring at decreasing values of **I** as μ decreases. It is also important to notice that as values of **I** increase, values of population/preference-information and population/behavior-information converge. This is not surprising because the selection mechanisms for the π distribution are borrowed from the β distribution and, as noted earlier, the consistent inequality in information between the two distributions is mostly due to the weaker selection on the π distribution. Again as before, both individual/preference-information and individual/behavior-information covary with genetic fitness.

Information and Selection

In contrast with the selection-only results of Chapter 5, in populations with variation it is not the case that higher population information necessarily covaries with higher genetic fitness, especially information in distributions of immediate behaviors. The point made earlier regarding the population/behavior-information of the simple bees holds also in the case of single distribution (or "single level") systems. Once one sees that higher mutation rates can increase the efficiency of environmental tracking while reducing fitness at equilibrium, it becomes clear that higher correlation of distributions to environmental states is compatible with lower average fitness over time. Information needs not only to be present, but to be exploitable as well. The process of interest here is one by which information is made exploitable.

Information held in the probabilities of individual responses, rather than the population distribution of such responses, did covary with fitness in all of these trials, both in distributions of preferences and overt foraging behaviors. Reflecting back on the discussion of the relationship between payoffs and information in Chapter 4, it is not hard to see why this is the case. Increasing the environment-response covariation of already-adapted responses is likely to increase payoffs, especially in situations with only two state-response pairs and where the payoffs are symmetrical. This accounts for the perfect covariance between individual/behavior-information and fitness in these trials, and we should expect to see some failures of covariance in systems with asymmetrical payoffs or more than two state-response pairs. It is of some epistemological interest that the population-level information in preferences, although lower than that in behaviors, covaries with fitness as well. The relationship of population information to fitness is more complex and, as such, of more technical interest.

Recall that in the simple bees, population information was high and individual information was low, as was fitness. The population as a whole seemed to be good at keeping track of the state of the environment but was unable to fully exploit that information. Thus, population-level information held in behavior distributions fails to covary with fitness under a variety of parameter changes.

The preference subsystem of the smart bees coopted or "exapted" the sensor mechanism which accounted for the high population-level information in the simple bees as a selection mechanism in the dynamics of an internal "preference" mechanism. This exaptation of a success indicator into a new role had a number of interesting effects. Oddly, population/behavior-information went down, because the introduction of foraging bias reduced population-tracking efficiency. Moreover, instances of near-perfect population/behavior-information were an indication of the breakdown of the preference/behavior-control mechanism. The upside is that both genetic fitness and individual/behavior-information went up.

The addition of the preference level added a new kind of information, which is significant in our fumbling toward a model adequate to the understanding of human knowledge. Individual/preference-information (analogous to beliefs or desires) covaried with genetic fitness, which is to say, with states of the world relevant to fitness. Population/preference-information (analogous to the state of culture) covaried under conditions of fixed mutation rates on preferences, and although the connection with fitness was not as strong as individual information's, population-level information was generally greater.

What covariance of informational quantities with fitness implies is that for these kinds of systems, when natural selection on hereditary traits relevant to this kind of control structure increases reproductive fitness, they increase information on a variety of levels, as well as selecting mechanisms which exapt existing sensors to make their information more exploitable. So, not only is there strong indication that selection can drive information increases and information processing sophistication, but we also have a rigorous model with which we can begin to investigate the principles of this process of selection and information gain.[9] Or, more prosaically, selection results in internal states becoming better indicators of external states, driving the optimization and elaboration of knowledge systems. Of course, this is just what evolutionary epistemologists always assume is the case, but the trick is being able to build a modeling framework within which the relevant research can proceed, while maintaining the requisite ontological neutrality. I like to think that the current model constitutes a good start on this project.

Information and Meaning in Evolutionary Processes

IMPLICATIONS FOR NATURALIZING EPISTEMOLOGY

One of the more remarkable things about science is how much of the real work is in proving things that are already known or strongly suspected, in quantifying relationships *gratuitously,* before there is any practical reason to do so, and in wantonly simplifying the complex. People knew long before Newton that what goes up must come down and that the farther it falls, the faster it falls. But the analytic geometry developed in the seventeenth century allowed the quantification of movement, while demanding the abstraction of complex physical solids to mere point-masses. It was this theoretical framework that allowed the insight that, when the apple falls, *the earth moves also,* an insight which allowed the unification of the theories of terrestrial and celestial motion.

The epistemological problem which we have been trying to get a handle on for the last five chapters is a lot like gravity, in that people take their knowledge of the world for granted and are reasonable in doing so even though they have no theoretical understanding of the nature of the knowledge relationship nor of why it is they are reasonable in taking it for granted. Like gravity, it seems rather unlikely that, at least in the short term, a theory of knowledge is likely to make any real difference to the reliability of knowledge. Unlike gravity, however, the lack of a coherent scientific understanding of knowledge is a particular embarrassment, since science claims to *know* so many things and has acknowledged ambitions toward the construction of a systematic, coherent, and all-encompassing worldview.

What we have been involved in for the last five chapters consists of the staples of scientific theory building: wanton simplification, gratuitous quantification, and the attempt to prove what everyone already knows – that their internal states reliably track their changing environment in useful ways. Few, of course, hope for Newton's kind of success, and it seems to me that, rather than the framework developed here constituting the *foundations* for a proper quantitative naturalistic epistemology, the mathematical models of evolution, selection, and information are but a few of the tools necessary to take epistemology from being an introspective exercise to its integration with the larger body of theoretical and empirical investigation. Nonetheless, the concepts of selection and information, as well as their relationship in both single and multiple population systems, are critical to bringing some of the distinctively philosophical problems of knowledge within the scope of the broader project of the evolution of cognition and learning.

Let us pause and review. The project began with several intuitions and convictions: the acceptance of the appearance-reality distinction (the

"Kantian barrier") as the basic problem for epistemology. The idea that possibly the only thing we can confidently say about the world as it is in itself is that it must be such that we do in fact experience the things we do. It is not simply my experience that places constraints on how the world can be, but that the facts of my *existence* and *persistence* reveal something about the world as well. The facts of my existence, persistence, and experience alone, however, are in themselves insufficient to further the foundationalist pursuit of certainty. We are not after certainty, however, and despite the reductionist architecture preferred herein, the ultimate test is always one of coherence and austerity of assumptions when opening up new areas of inquiry. So, we add a carefully chosen few extra assumptions: that the functional complexity we observe in ourselves is the product of the history of our interaction with the world, both in our own lives and in our lineages – that is, the product of evolution by natural selection of heritable variations. It turns out that, within the basic epistemological situation, evolution is a remarkably productive assumption, even in its most austere or ontologically neutral guise. For not only does it allow us to address the issue of the functional reliability of our senses and cognitive mechanisms without cheating by "peeking behind the curtain," it also allows us to understand how evolved control systems like ourselves *impose* distinctions on the world via the suites of options we present to it. In short, we began with the intuition that natural selection, or differences in real stabilities of various sort of things, transfers information about the world across the Kantian barrier. The challenge has been to capture this intuition and put it to work.

Chapters 3 and 4 were dedicated to the examination of basic evolutionary and informational concepts with an eye toward (1) making them as clear as possible, (2) giving them the mathematical characterizations necessary for systematic study, and (3) stripping them of assumptions that impede the epistemological project. Fortunately for us, it turned out that existing mathematical characterizations of evolution and information, like statistics, are indifferent as to the domain of their application. In other words, our mathematical theories already have the requisite ontological neutrality to support an evolutionary epistemology.

Chapter 5 examined information transfer from environments to population distributions for single populations and established that there is a wide variety of ways we can characterize environments simply in terms of the stability properties of individual types (fitnesses) and partitions on the space of the fitness vector $\vec{w}$. The proof (see Appendix) established a constant momentary increase in information under selection alone, although this should be understood as one of several competing tendencies in evolving systems (fitness fluctuations and variation both counteract this tendency).

Chapter 6 introduced the notion of a fitness feedback loop between genetic and transient-behavioral distributions – essentially a formal model of the behavior of Campbell's "vicarious selectors." The results of Chapter 5 indicate that each population will, under favorable circumstances, accumulate and maintain exploitable information about its local environment, factors that drive its own evolution. The model in Chapter 6 shows how in such systems we can also expect behavior distributions to maintain information about certain aspects of the *genetic* environment; in particular, about which of the available behavioral options is locally or currently best for genetic fitness. In this we see a relationship that begins to resemble our own epistemological situation. Chapter 5 simply suggests that which behaviors I exhibit indicates which behaviors are most stable. Chapter 6 shows why I should expect my behaviors to be usefully and reliably coordinated with those factors which have historically affected the stability of the inherited part of my constitution. Chapter 7 elaborates the model one step further, showing how this same sort of information also ends up in internal states and how models of this sort provide us with a wealth of precise concepts of information, which – along with the open-ended nature of the population network model – allow the asking of many interesting and precise questions about how the evolution of cognitive architectures results in new kinds of information about the world being maintained in cognitive systems.

So, what has been established? That the simple fact of evolution, even in its mildest (and thus most broadly applicable form), is sufficient to open up to systematic study the relationship between appearances and reality. Or, rather, because we understand that reality remains as representationally remote and yet (causally?) immediate as it always has been, what evolution opens up is actually just the *comparative* study of adapted cognitive relationships to the world. The tools developed are designed to be scalable in the sense that, while the models built here only establish certain principles relevant to the human epistemological condition, the multilevel selection framework can be elaborated to accommodate more distinctive features of human cognition, such as the belief-desire distinction and subject-predicate representation. But by beginning the analysis in this way, without characterizing the adaptive targets of simple cognition in our own terms, when we arrive at a model of human cognition we may finally be able to say what it is in the world that *this* thought is about without employing synonyms ("cat" refers to felines) and thus without presupposing that we have somehow got it right. Once we can describe our own thoughts without presupposing that they are the "end of the road," perhaps we will be in a position to do a little productive engineering of our systems of knowledge.

The approach to modeling information transfer in evolutionary processes taken herein is obviously not the only one possible, and something should be said regarding what features of this model are critical to its success and prospects.

Reification: Ontological neutrality is critical for any interesting epistemological project, and the key is not to *reify* what lies on the world side of the barrier. Avoiding reification is simple, one just uses tools that map states rather than objects. Mathematical models commonly have this property, and this is why population dynamics and mutual information are suitable tools for evolutionary epistemology. Reifying what lies on the observable side of the barrier is less problematic, but moves away from objects and toward states increase generality and austerity.

Probabilities: Tools of probabilistic analysis like fitness and entropy are essential because the regularities of nature we (hope to) track as well as any systematic response to them are statistical in nature. Population-level models (either vector models like the ones here or agent-based models with large populations) play a critical role here since they provide a basis for assessing probabilities, with ideal or effectively infinite populations maximizing the accuracy of assessment.

Multilevel Models: Any successful evolutionary epistemology must allow us to investigate the interaction between the evolution of organisms in generation time and the patterns of phenotypic variation that constitute development, learning, and communication.

Selection and Variation: It is not necessary to follow Campbell in insisting that we have a common concept set for analyzing the dynamics at every level. Certain processes like probabilistic learning are awkward to model in terms of selection and variation, and we should expect that as we approach adequate models of human cognition, we will find it inconvenient to shoehorn every dynamic process into the replicator dynamics. As a preliminary tool set, however, the generality of selection and variation as developed in Chapter 3 recommends it. Moreover, modeling cognitive and cultural processes in terms of variation and selection (trial and error) provides one way of understanding how genetic constraints allow open-ended creativity while maintaining at least minimal functionality.

Mutual Information: Mutual information was introduced herein mostly to give us something precise to talk about when talking about information, as well as providing a statewise measure of tracking efficiency that is formally independent of fitnesses. Presumably, other measures could be substituted or, indeed, the project could proceed without an information measure. The results from Chapter 7, however, indicate that an information measure is needed to spot adaptive opportunities, such as when population-level information is high and individual information is low.

In short, evolutionary epistemology requires mathematics rather than metaphors and a framework for the analysis of interactive multilevel dynamics. The preliminary construction herein shows, I think, that the project is possible and that the comparative scientific study of cognitive systems can, in fact, address one of the central problems of epistemology. Other problems remain, however. The most important have to do with the nature of the apparently nonfactual normative claims we make when we consider whether a belief is true or if we are justified in having a belief. This, I will suggest, should be approached by asking what such statements *mean,* and that the appropriate theory of meaning is sufficient to answer criticisms to the effect that naturalistic epistemology must treat such statements as mysterious. The deployment of such a theory of meaning is the purpose of Part III.

III

Meaning Conventions and Normativity

8

Primitive Content

Many philosophers who are sympathetic to the kind of mathematical evolutionary approach to the general problem of state individuation and environmental coordination developed in Part II will nonetheless contend that it fails as an adequate comprehensive approach to epistemology. For epistemology, according to tradition, is, like ethics, a *normative* discipline. This means that it is concerned not so much with how people form beliefs, but how they *ought* to form them. According to the traditional conception, its objective is a normative theory – that is, a theory which goes beyond merely describing the system of rules for knowledge acquisition to generate authoritative pronouncements regarding how one ought to form beliefs. The laws of reason and evidential support are its natural subject matter, the articulation and defense of scientific method its ultimate goal.

It is often said that the problem with naturalistic approaches to the study of knowledge (like the one taken in this book) is that by taking on the mantle of science one foregoes the ability to make any pronouncement regarding how anything ought to be done. Science deals with the facts, and the facts, as we know, are supposed to be value-neutral; the scientific approach to knowledge can thus never tell us anything about how we ought to form our beliefs; naturalistic epistemology is not epistemology at all because it *cannot* be normative. Thus, one can concern oneself with how people form beliefs, with how they ought to, and with whether people in fact form beliefs the way they ought to, but you *cannot* go from how people do form their beliefs to how they ought to for the simple reason that because something is so doesn't mean it *ought* to be.

The classic response from the original naturalistic epistemologist, W. V. O. Quine (1969), insists that we simply abandon the normative mission. Epistemology is to become a branch of psychology, mapping organismal responses in various environments. As always, there are lots of other reactions, and the interested reader is directed to Kornblith's (1994) anthology of classic papers, *Naturalizing Epistemology*. Most philosophers consider Quine's position rather extreme, however, and even for those of us who never quite understood the motivation for normative theory building – systematizing ought-statements – the total abandonment of the traditional concern with the normative seems to leave a bit too much baby in the bath water. Which is to say, even if one has no aspirations toward telling people what they ought to do or how they ought to form their beliefs, it may still seem that understanding normative concepts such as truth and justification is just as central to a philosophically adequate understanding of human knowledge as understanding the tracking relationship between thoughts and things-in-themselves, which I characterized in Part II as the central problem of epistemology.

It would seem, then, that even short of dispensing the sort of advice that normative theorists intend, there remains for the naturalistic epistemologist either some deep mysteries or some distasteful simplifications. For either one must accept the normative authority of the "laws" of reason and standards of evidential support more or less at face value and try to reconcile such acceptance with one's naturalistic scruples, or one must reject them entirely along with ghosts, witches, and psychic "energy" as being ungrounded in the kind of observational confirmation requisite for items of scientific ontology. The mystery comes in because if meaning and normativity do exist in physical terms, we have no idea what they are. If not, then we must dismiss as unfounded much of what people seem to find most important in life.

The next two chapters are devoted to explaining why I think naturalism will ultimately dispel the mystery without giving up the goods. I do not expect to convince any but the most congenial reader, simply because I am not going to write enough pages to even begin the point-by-point rebuttal of the hoary tradition which maintains the utter incomprehensibility of the normative from the scientific point of view, that "is" and "ought" are two wholly dissimilar sorts of things and that it takes a *mind* to imbue a sign with meaning. Fortunately, the theory I have in mind can be explained fairly concisely. What I have to say is divided into two parts – theory in this chapter, defense in the next. This means that my comments on such philosophical mainstays as the arguments of Hume and Moore will have to wait until the next chapter, but just so the reader has some idea where all this is going, I will say something about my general position here. I do not believe that ought-statements are *derivable* from is-statements, and I think I can explain why that is so in naturalistic terms. I do not think that science has normative authority in the sense of being able to say what we ought to do or want. I do think that signs have meaning without minds (whatever those may be) and that the correct theory of meaning allows us to understand the relationship between "is" and "ought" – and, consequently, to say *in principle* what it is that makes statements about truth and warrant true. I emphatically have no intention whatsoever of *answering* the question "what is it rational for me to believe?" I am quite interested in the question, however, and in particular in what would make an answer to the question *true*.

Finally, although it will become clear that I believe what I have to say applies to ethical norms as well as to epistemic norms, I have no intention of defending a moral theory, for two reasons. First, the demands of social organization require different sorts of flexibility than those required of factual representation; thus, the complexities of moral authority need to accommodate much that is unique to morality. The epistemological model to be presented no more than suggests how the corresponding model of morality would look. Second, addressing the full wealth of recent trends in the theoretical foundations of ethics is simply beyond the scope of the current project (and my expertise). Unfortunately, it will not be possible to forestall all mention of ethics because most of the arguments against naturalized epistemology are simply gestures toward long-standing arguments about the limits of naturalism in ethics. If my criticisms of such arguments seem to undermine them on their home turf, this should nonetheless be understood as a defense of evolutionary epistemology rather than a foray into metaethics.

I suggest that the key to dispelling the mystery of intentionality and the mystery of the is-ought gap lies in a functional theory of meaning of the sort developed by Millikan (1984). Rather than asking what kind of arrangements of the world could have normative force for rational humans, one considers in general the adaptive function of a language or signaling system and then asks under what conditions a signal or statement fulfills that function. There turn out to be as many kinds of meaning as there are kinds of signaling systems. Contemporary theories of meaning have typically gone astray (taking much of the rest of analytic philosophy with them) in assuming that there is only one kind of meaning or truth-bearing entity – the proposition. Propositional structure involves the conjunction of subject and predicate, which is just the attribution of a property to an object. But not all systems of signals work the way statements of fact work, and the truth (or satisfaction of tracking conventions) of signals depends on *how* the systems they are embedded in fulfill their evolutionary design. Even more critically for our purposes, it is the function of a signaling system that combines different aspects of meaning in the same signal. Unlike the indicative language of science and rational thought which is usually taken as the paradigm for truth-bearing signals (and the model for the proposition), most natural signaling systems motivate directly just as normative language seems to. Understanding the difference between "is" and "ought" is just a matter of understanding what is-statements and ought-statements mean, and their various meanings come from the different functions of is-statements and ought-statements and the internal control systems that issue them. It turns out that all we require is a fairly broad understanding of how meaning conventions arise from the history of adaptive or stabilizing functioning in signaling and control systems.

The way through the mystery, then, is to ask not what properties of things have normative force, assuming that our common language always tracks the world in the same way, but to ask what it would take for a statement directly regulating behavior to be true – what are the tracking conventions for normative statements? More to the point, what are the tracking conventions for normative *intuitions*? The reason I emphasize this is because it seems to me that what we need is not so much an analysis of the meaning of the words we use when we make normative statements as an analysis of the meaning of the *feelings* of wrongness which we express in such language. Such an analysis requires some idea of what sorts of *rules* of meaning apply to languages and signaling systems and what kinds of jobs normative sentiments or intuitions have. The rules we are going to be looking at are the rules or conventions that emerge from the history of adaptive design, although I should point out that for the most part, any other account of fallible function can be used in a similar manner to explain the meaning of ought-statements.

MEANING CONVENTIONS

Oddly, the place to start in understanding the functional basis of meaning is not with the word, which may be a peculiarly human invention.[1] Words by themselves have no function. Only in the context of complete sentences do they have a job to do. "Book" and "red" are neither true nor false for they make no claim. They do participate in the tracking conventions of our language, but it is not until they are part of a complete sentence that we can evaluate the aptness of their usage. For instance, "book" does not always require the presence of a book, for we can say "this is not a book." The meaning of words turns out to be rather more complicated than one would expect, and they are certainly not the proper analogs to the signals sent by animals. A complete sentence, on the other hand, is a signal with a job to do and conventions governing the conditions under which it fulfills the tracking conventions of the language as well as conventions governing what follows from it either in terms of direct behavior, in the case of commands, or what

Primitive Content

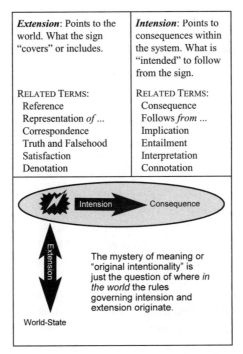

Figure 8.1. Conventional elements of meaning.

other sentences follow from it, in the case of statements of fact. The weighty, implication-laden statement that I will simply let fall before moving on is that *the word is not the basic unit of meaning*. The supposition that the word *is* the basic unit of meaning is responsible for numerous mysteries, including the relationship between "is" and "ought," the problem of what feelings or emotions mean, and the problem of what it is that animals might be communicating to one another. Even sentences, despite their representational completeness, are not the proper analog to animal signals – but let us return to this subject later.

The thing that makes meaning or semantics so peculiar from the point of view of science is the directedness or "intentionality" of meaningful representations. Each representation is directed toward or "about" at least two sorts of things (Figure 8.1). First is the thing that the representation is a representation *of* what it stands for, what it "corresponds" to, or what makes it true. When talking about the meaning of individual words (component parts of representations), this is usually called the reference or, more technically, the "extension" of the representation. We will use the term here for complete representations as well. In the case of complete sentences, this makes "extension" equivalent to the "truth-conditions" of the sentence. The second part of meaning, and we often take it for granted, is the nest of relationships governing the internal functioning of signs and signals in a language or representational system. In human language, these include the definitions of terms (which are often taken to determine their extensions), the logical implications of sentences, the "modes of presentation" (like attributing beliefs rather than expressing them), and various attitudes one can have toward propositions (e.g., believing that *p*, hoping that *p*), which together weave the collection of signs and symbols into a representational *system*. We will adopt the term *intension*, which is used (irregularly) to refer to various of these internal components of meaning, to refer to the totality of system-internal meaning relationships, as intension is often used as a complementary term to *extension*. Both of these relationships – extension and intension, reference and implication, call them what you will – are intentional (with a "t") relationships. This is to say, they are not causal or, at least, not occurrently causal. There is instead some sort of rule or convention that applies, which for extension specifies *when* the representation is supposed to occur or what state of affairs it stands for, and for intension specifies what one is supposed to make of the representation, what follows from it, or what it implies. At various times and for various people, intentionality of both sorts, what a representation means in the sense of what it stands for and what it means in the sense of what follows from it, have been taken to be the distinctive mark of the mental. The conjunction of meanings of both sorts, a reference and a collection of entailments, constitute what is called the *content* of a representation or mental state.

CONVERGING ON A NEW THEORY OF MEANING

One of the more interesting developments in the philosophical study of meaning in the last twenty years is the convergence of two approaches to the understanding of meaning on a single architecture. The two approaches I have in mind are Millikan's teleosemantics (1984, 1993) and a formal model of the evolution of meaning conventions by philosopher and game theorist Brian Skyrms in his *Evolution of the Social Contract* (1996). Although both Skyrms and Millikan attempt selection analyses of meaning, their starting points differ considerably, as do their methods. Millikan's construction is considerably more elaborate as well. What juxtaposing her treatment with Skyrms's does is highlight a certain basic kind of meaning, which I am going to suggest is the key to understanding the difference between "is" and "ought" and, consequently, to allowing naturalistic epistemology to account for if not generate the normative elements of human knowledge.

The place to begin is with the adapted "signaling system," a term Skyrms borrows from Lewis's *Convention* (1969). Following Lewis, Skyrms asks us to consider a cooperative game with two players in which one player is designated the sender (S) and the other the receiver (R). In the simplest case, there are two world-states (T1 and T2), two messages (M1 and M2), and two actions (A1 and A2). A1 is designated the correct action, in T1 and A2 in T2. If the receiver performs the correct action, then both players get a payoff of 1 point, and otherwise nothing. The problem is that only the sender can detect the state of the world. To make matters worse, we suppose that (despite the suggestive numerals) the available signals M1 and M2 fail to resemble in any way either world-states or actions but are, nonetheless, the only way that the two players have to communicate. Needless to say, this is a game of pure coordination (both players getting the same payoff at the same time). All that is necessary is for sender and receiver to coordinate their sending and receiving strategies in order to get the payoff every time.

In good game-theoretic fashion, we consider all the possible strategies for sender and receiver. There are four sender strategies and four receiver strategies, a follows:

Sender Strategies	Receiver Strategies
S1: Send M1 if T1; M2 if T2	R1: Do A1 if M1; A2 if M2
S2: Send M2 if T1; M1 if T2	R2: Do A2 if M1; A1 if M2
S3: Send M1 if T1 or T2	R3: Do A1 if M1 or M2
S4: Send M2 if T1 or T2	R4: Do A2 if M1 or M2

We further suppose that individuals play the game repeatedly and at each round can either be sender or receiver, more or less as one would expect for animals sending danger calls and the like. Each player, therefore, needs a combined strategy which consists of a sender strategy and a receiver strategy. There are sixteen such strategies, which we can designate S1R1, S1R2, S1R3, and so forth.

What Lewis had in mind in designing this game was to give an account of meaning based on the conventions that functionally coordinate senders and receivers. In Lewis's technical sense, a "signaling system" is a complete strategy which can optimize payoffs. In the simple two-state example, there are two such strategies: S1R1 and S2R2. These two strategies always manage to get the receiver performing A1 in T1 and A2 in T2 for the full payoff to both players. The two strategies use different signals for the two situations, however.

S1R1	S2R2
T1 → M1 → A1	T1 → M2 → A1
T2 → M2 → A2	T1 → M1 → A2

Thus, the meaning of the signals M1 and M2 is conventional, where the conventions emerge somehow from whatever process governs the stabilization of the strategy. For Lewis, the idea was that rational agents with common knowledge of each other's rationality and understanding of the game would each arrive at a collective strategy. That collective strategy would be rationally optimal, and this would yield a rational basis for meaning conventions. Skyrms raises two problems with Lewis's approach. First, even ideally rational agents with the required common knowledge require some sort of initial inclination toward using the signals one way or the other. Unfortunately, the situation is too symmetrical for idealized decision makers to sway one way or the other. While such inclinations are natural in the real world, they are not part of the concept of ideal rationality. Skyrms has a deeper criticism as well. If the rational-choice game is supposed to account for the basis of meaning, then one cannot begin by assuming idealizations such as rational agents who are *already* fully capable of thinking about each other's thought processes.

Skyrms's solution to both of these problems is to turn the game for rational agents into an evolutionary game, in which strategies are inherited rather than chosen and competition between strategies occurs in the real world rather than inside the minds of rational agents. The real-world aspect of the new situation solves the symmetry problem – the noise inherent in evolutionary systems will inevitably provide an inclination in some direction or other. More important, the choice between strategies is made by the environment via natural selection rather than by minds already imbued with meaning. Selection causes the stabilization of one set of conventions or the other, providing a solution to the game that Lewis could not quite reach. We get the emergence of meaning conventions in a way that does not require their previous existence.[2]

Skyrms goes on to discuss a number of computer simulations which indicate that in open competition with the full set of strategies, one of the

Primitive Content

two signaling systems inevitably win. This is not surprising, considering that twelve of the sixteen strategies involve either a sender or receiver who always does the same thing. The remaining two are S1R2 and S2R1 – "antisignaling" strategies which manage to get the receiver always doing the wrong thing. The important point for Skyrms is his general theme that evolution can solve problems that even ideally rational agents cannot.

For us, Skyrms's evolutionizing of Lewis's signaling game has another significance, one that lies in the architecture of the signaling systems rather than in the fact that evolution can choose them when reason cannot. For Lewis, sender and receiver strategies were paired by rational agents in the process of choosing the ideal strategy to play with each other in complementary roles. In Skyrms's version, strategy pairs are chosen by the environment. If we take the evolutionary analysis further, supposing that sources of variation such as mutation and recombination operate on the physical structure which actually implement the strategies, we see that sender and receiver are adapted to the way the world works in offering perceptual stimulus and preferring action, as well as to the manner in which the complementary part is adapted to the world. Sender and receiver coevolve and become coadapted. The meaning of the signals derives from the conventions that facilitate the coordination between sender and receiver which allows this. So, for example, M1 means-extensionally T1 and means-intensionally A1 in a community in which S1R1 has risen to dominance through natural selection.

What Lewis and Skyrms fastened on as a basic model for studying the conventional basis of meaning has precisely the structure that Millikan begins with in her evolutionary analysis of meaning. The terminology differs, of course, Millikan focusing on the selected *function* of a representational system rather on conventions per se. Millikan was also laying the groundwork for a much more elaborate model capable of dealing with grammatical syntax and word meaning as well as with the traditional puzzles that concern them, so she needed to be fussier about terminology. What distinguishes a "proper" function from the general class of processes, however, is its adaptive history – the manner in which the process, trait, or system has made its adaptive contribution.

To cast things in Millikan's terms, a representation[3] (the signal) mediates between coadapted producers (senders) and consumers (receivers) of the representations. The truth-conditions for the signal are just those "normal" conditions in which the system has made fitness contributions to the organism, what we have been calling the extension of the signal. The proper function of the producer is to get the representation sent in those conditions, although most often this is a matter of responding to some indicator of those conditions. The proper function of the consumer is to respond to the representation in some particular way, determining what we have been calling the signal's "intension." Despite the fact that Skyrms conceives of sender and receiver as part of a whole and Millikan emphasizes the separateness of producer and consumer, the systems are isomorphic. It is the matching of signal production and consumption processes under selection that determines the meaning of signals.

An example will be helpful at this point. Cheney and Seyfarth (1990) discuss a signaling system in vervet monkeys in Kenya. It seems that vervets must cope with three kinds of predators: pythons, eagles, and leopards. Vervet sentries issue three kinds of warning cries and hearers engage in three kinds of evasive strategies. The cry for eagles causes vervets to look up into the sky. The cry for pythons causes them to stand up and look around on the ground. The cry for leopards causes them to run up the nearest tree. The parts of this signaling system are designed to work together. The vervets' perceptual system must result in the issuance of the right cry for the circumstances. What makes it the right cry depends on what the vervets are supposed to do in response to the cries. The design of the signaling system is such that there are rules for the issuance of signals and rules for responding to those signals. The two sets of rules must fit together as part of a unified design; otherwise, the system as a whole does not function. Skyrms's model emphasizes that individual monkeys must be prepared to play either role, sentry or citizen. Millikan's emphasizes that, strictly speaking, different faculties – neural structures – are involved in sending and responding to the danger signals. Sending and receiving systems are coadapted because they are (presumably) free to vary independently under selection. The two analyses are fundamentally analyses for the same simple sort of signaling systems. Neither assumes words, combinitorial syntax, or even the separation of environment-indicating functions (e.g., beliefs) from motivating functions (e.g., desires informed by beliefs). Rather, both assume that the most basic vehicle of meaning is a sort of monolithic signal which both tracks environmental states and motivates directly. Just as in Chapter 2, where we considered whether the cell might be the "atom" – the proper explanatory nexus – of evolutionary biology, here we consider whether Millikan's "intentional icon," animal communications like the vervet's warning cry, might be the proper focus of a naturalistic understanding of meaning.

Figure 8.2 provides a schematic diagram of the functional architecture of signaling systems. Borrowing Millikan's terms, representations (the lightning bolt) mediate the coadapted function of a representation producer and a

Primitive Content

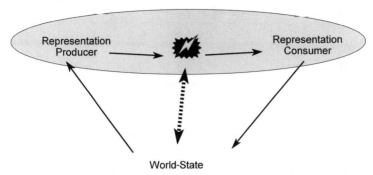

Figure 8.2. The causal loop of a functioning representation system interacting with the environment. The solid arrows indicate causal processes, and the dotted arrow indicates the correspondence relationship whose satisfaction determines truth-value.

representation consumer. In the case of individuals, the producers are perceptual apparatuses, the consumers behavioral control apparatuses, and the representation is the signal that mediates between the two, allowing the behavior to get performed in situations where it is beneficial. The solid arrows indicate causal processes and the dotted arrow indicates the correspondence relationship, which is supposed to be maintained between the signal and the varying states of the world and whose satisfaction constitutes truth for simple systems like these. Just as in proper languages, truth is the satisfaction of the tracking conventions that apply.

Meaning is conventional, and conventions are historical entities. Much of the ink spilled on the topic of meaning could have been saved if this simple fact were more often repeated. The pertinent question is how the relevant conventions emerge. The natural tendency when looking at the meaning conventions that emerge from the function of a signaling system is to focus on the actual causal pathways involved in the production and consumption of the signals and to assume that the reference of the signal must be determined by the production conventions. Dretske's (1986) indicator semantics, as discussed in Chapter 4, has this common feature. The reference of the signal was to be determined somehow by the *causes* of perception under historically adaptive conditions. The "sense-data" theories of the early-twentieth-century empiricist shared this assumption, as is evident in Quine's term "stimulus meaning." The stimulus meaning of a perception was to be the (statistically) normal cause of the perception. Such "supply-side" approaches to naturalistic semantics have been plagued by difficulties in picking out referents, a subject touched on in Chapter 4. The problem is that supply-side specifications of a signal's extension still include too many possible referents. For epistemological purposes, there is no such thing as "the cause" of a perception. Rather, there is a whole chain of things which result in a perception. Too many things fit the bill. Consequently, the obvious suggestion that a signal is about its cause never gets off the ground.

Both the production and consumption of signals are important to understanding the epistemology of a signal. Among contemporary philosophical meaning theorists, Millikan may be unique in insisting that the extension or truth-conditions for the signal have relatively little to do with the conventions governing the production of the signal.[4] For instance, the state of affairs that constitutes the truth conditions of the vervets' leopard cry is just that state of affairs in which it has been advantageous to run up a tree in response. Given the vervets' history, this is just when there is a leopard close enough to pose a danger. The reason that the cry means "leopard" is that the signaling system is designed to get the cry to covary with the presence of leopards. Of course, limitations on perceptual response do place limits on those design possibilities and, thus, play an indirect role in determining extension.

PRIMITIVE CONTENT

Notice that in simple signaling systems, it is not at all clear whether the signals are indicatives or imperatives. Despite the fact that they seem to possess conventional rules governing extension and intension (insofar as they have adapted functions), they defy any sort of easy categorization.[5] Why is this? By which I mean not why are they hard to categorize, but why do we find them so hard to categorize, despite their absolute ubiquity? The answer, again, is that we have been approaching meaning anthropocentrically, attempting to subsume animal signals into the human framework rather than the other way around. What happens when we start with animals and work our way toward human beings, as a good biological naturalist should?

Let us suppose, then, that Figure 8.2 gives us a picture of meaning that is biologically basic. Signaling systems arise serving communication both between and within organisms. Whether warning cries or scientific speech, neural impulses or activating hormones, in each case a signaling system requires two cooperating devices, a sender and a receiver, a producer and a consumer, who are jointly selected on the basis of some process that is initiated by a signal. The signal is sent by the producer in response to some stimulus. The signal elicits in the consumer some action which has consequences affecting system stability. Conventions exist governing which signal is to be used just in case the signal used for a particular purpose is conventional, which is to say, it could have been otherwise. As the Lewis-Skyrms game and Millikan's schematic of a representational system bring to the fore, two sorts of conventions apply: (extensional) tracking conventions and (intensional) consequence conventions. In simple systems, the extension of the signal is not specified in subject-predicate terms, but via the conditions of efficacy of the behavior following from the signal. The intension of the signal just is that behavior. How does this compare to the rational thought of human beings?

Human communication and decision making has rather obviously more "moving parts" than the analogous systems of most other animals but, in either case, the overall task of the system is to get behavior profitably coordinated with circumstances. The tracking-and-motivating signals we have been considering do this all in one step. Human decision making, on the other hand, is removed from this in two important ways. First, rational behavior is not automatic upon receipt of a tracking signal but is always conditional upon the individual's desires or perceived needs. The standard theory of rational action characterizes this as the distinction between belief and desire. According to this model, without desire, beliefs have no behavioral consequences. Without beliefs, there is no way to decide how to go about attempting to fulfill desires. This distinction is so basic to human thought that the essence of the standard theory of *action* is often captured in this simple formula: "belief plus desire equals action." Rationality itself is usually defined in these terms as well: the rational agent acts so as to maximize the expected satisfaction of desires, given his or her beliefs.

It would be a mistake to assume that the belief-desire division is a uniquely human construction. On the contrary, it seems evident that most animals in their internal processes divide tasks up in a similar way. Foraging behavior would almost have to be controlled this way. The presence of food or prey (as determined by external sensors) combined with the need for food (as determined by internal sensors) results in hunting or feeding. Thus, in human and animal systems alike, the all-in-one representational functionality of the danger signal is broken down into two independent parts, roughly corresponding to the indication of internal and external states. Action or behavior results from the joint occurrence of two sorts of occurrences. Thus, behavior is no longer automatic upon the indication of an external state but is conditional on current internal states as well.

Second, as noted earlier, in human thought and communication, indicative sentences can be further broken down into word elements. Words combine to create sentences indicating environmental states, whereupon sentences combine with perceived needs to motivate actions.[6] All of these parts, all of this structure, is required to fulfill the same function as the single tracking-and-motivating signal – coordinating behavior with environmental states – which is the overall function of every representational system. The simple tracking-and-motivating signal is semantically basic not because it has functionally the smallest job to do, but because the rules governing it are the simplest.

There are, doubtless, almost immediate advantages (as well as costs) to be had by making a representational system more complex – from condition-alizing behavior on both internal and external sensors to creating a scheme that allows associative learning about regular features and conditions of the environment. But each element of such a more complex representation system derives its meaning from its role in achieving the basic behavior coordinating purpose of every representational system. Simple signals track and motivate. Sentences indicate states, perceived needs determine ends. Names orient us to regular features of the environment. Adjectives condition our expectations, connectives allow increased precision in the specification of relationships, and so on. In each case, the meaning of representing elements can be understood with respect to the stabilizing conventions which apply to the system. Millikan's work goes a long way toward making this clear.

FORMALIZATION

What has made normative statements hard to understand is the fact that they appear to possess a kind of meaning involving rules governing both tracking and motivation. What the preceding discussion indicates is that the conjunction of tracking and motivation apparent in normative statements, rather than being unusual, constitutes the most basic kind of semantic content, a kind of content which arises in the simplest model of conventional meaning one can construct. Our confusion arises because our paradigm of meaning, the indicative sentence composed of referring words, is highly complex in one specific way, optimized for something like associative learning. In addition to making it difficult to see how one signal could both track and motivate, it makes it hard to see other kinds of semantic sophistication. Normative statements may resemble tracking-and-motivating signals in that they are motivationally complete, but this does not mean that they are no more than danger signals. Indeed, there is some reason to think that they are issued by a special sort of functional subsystem which we express with familiar awkwardness via our common spoken language – namely, the function-stabilizing mechanism. To make the structure and semantic conventions of function-stabilizing mechanisms precise, we need to back up and create some basic notational tools for talking about meaning conventions in general.

Primitive Content

The most general thing we need is a way of specifying rules or conventions that respects our ontological scruples. What has allowed us to do this so far has been appealing to the causes of selection for some trait. We can continue to exploit this resource. As before, mathematical or, in this case, set-theoretical, tools allow us to avoid the rich connotations that come with ordinary language.

In the simplest terms, a rule says what is to happen when. A rule is just a map from conditions to processes. Maps of this sort are characterized set-theoretically as sets of ordered pairs, restricted so that each condition only maps on to one process. This is precisely the structure of a mathematical function. What we want here is to capture the conditions under which a trait has been selected, as well as what the trait has been doing that had positive selective value under those conditions. The reason for emphasizing that the rules are merely stabilizing *conventions* is to emphasize that the rules involved are implicit in the functional history of the system, rather than explicitly represented and followed.

The formalization looks like this: rule that applies to an adapted trait AT is just a map from conditions to processes.

$$\mathbf{R_{AT}} = \{\langle \mathbf{condition, process}\rangle |\ \mathbf{AT}\ \text{was selected for performing}\ \mathbf{process}\ \text{in}\ \mathbf{condition}\}$$

In case this notation is not familiar, curly brackets { } enclose the set, ordered pairs are enclosed in angle brackets $\langle x, y \rangle$, and the vertical bar | means "such that." The formal statement reads literally, "The rule for **AT** is the set of all ordered pairs with a condition in the first place and a process in the second *such that* **AT** was selected for performing the process in the conditions." This gives us a purely formal characterization of the rules we need without the risk of smuggling in the normativity we are trying to explain.

The rule (the set or ordered pairs we just defined) for a complex mechanism will contain various sorts of individual mappings. In each case, the processes are something that the mechanism can *do*. The conditions are more varied. They may be proximal causes of the specified process, in which the rule describes a causal chain of events. They may be states of the world which obtained when the processes were adaptively performed. This is important because functional correspondence maps involve the latter sort of condition.

The minimal architecture for generating primitive content is as follows: There is a signal-producing mechanism **P** which issues a set of signals $\mathbf{S} = \{\mathbf{s}_1, \ldots, \mathbf{s}_n\}$. The signals, in turn, elicit a set of responses $\mathbf{B} = \{\mathbf{b}_1, \ldots, \mathbf{b}_m\}$ from a response mechanism or signal consumer **C**. From the general rule governing an adapted signaling system, we can extract three kinds of subrules relevant to the present analysis. The rule governing the consumer includes specification of the *interpretation of signals*.

$$\mathbf{R}(\text{intension})_C = \{<\mathbf{s}, \mathbf{b}> \mid \mathbf{C} \text{ has been selected for } \mathbf{b}\text{-ing when } \mathbf{s} \text{ is received}\}$$

The rule governing the signal-producing mechanism **P** includes a correspondence map from states of the world ($\mathbf{W} = \{\mathbf{w}_1 \ldots \mathbf{w}_m\}$) to signals.

$$\mathbf{R}(\text{extension})_\mathbf{P} = \{<\mathbf{w}, \mathbf{s}> \mid \mathbf{P} \text{ was selected for sending } \mathbf{s} \text{ in } \mathbf{w}\}$$

Notice that the extension rule is different from the production rule for **P**.

$$\mathbf{R}(\text{production})_\mathbf{P} = \{<\mathbf{stimulus}, \mathbf{s}> \mid \mathbf{P} \text{ was selected for sending } \mathbf{s} \text{ in response to } \mathbf{stimulus}\}$$

The production rules will become relevant later when we consider the grounds for statements regarding epistemic *justification*. On the current account, these primitive versions of extension, intension, and justification are not normative in the sense that they are in human language but have the same form, which is to say that just because there is a historically determined rule which applies to some system doesn't mean there is anything intrinsically wrong with deviating from the rule. We shall return presently to the question of when violation of a rule might be wrong.

What exactly are the world-states **w**? Instead of the usual practice of specifying world-states by the attribution of properties to objects, we allow the functional architecture of the system to individuate world-states directly. Roughly, the extension of a signal is the *adaptive target* for signal timing. If an arrangement of signals and responses has been selected, then the history of the signaling system induces a set of partitions (another bit of set-theoretic apparatus) over states of the world, *whatever their ultimate nature*. Consider the simplest cases, such as the vervet or beaver warning systems in which each signal is designed to elicit one and only one response. The adaptive history of *each* signal-response pair carves the space of states of the world into two parts – those states in which selection has systematically favored the response (and which thus constitutes the adaptive target for perceptual discrimination and signal timing) and those in which it hasn't, the former constituting the truth conditions for the signal. The obvious worry here is whether signals pick up "anomalous" states as part of their extension, as when a false leopard cry causes one to avoid an unseen snake. The question to ask in such situations is this: what is the adaptive target for signal timing, given the nature of the environment and the way in which the signaling system facilitates adaptive coordination with it? The adaptive target of a signal cannot include situations with which it is impossible for the system to maintain productive correlation nor those with no consistent effect on the interests of the organism.

Unlike contemporary propositional semantics in which propositions are commonly interpreted as sets of *possible worlds,* the extension of signals in adapted signaling systems are sets of states of *this* world which have played a role in the system's adaptive history. Just as in Part II we were able to let behavioral efficacy determine the varying states of the world that biological knowledge systems track, here the same behavioral efficacy determines the extension of the signal, allows us to stay consistent with our ontological scruples. Perhaps, as realists propose, a good concept set aims to "carve nature at its joints," but nature has many joints that don't have immediate utility. Simple signaling systems pick the joints relevant to the suite of behavioral options they control, via the very efficacy of those behaviors.

REPRESENTING RULES

What we have so far is the formal expression of a broad and powerful theory of meaning. The remarkable thing about it is that it provides an infinite variety of nonpropositional contents, that is, combinations of intension and extension that cannot be expressed in an indicative statement. Such primitive content combines the tracking and motivating in a single signal just as our normative utterances seem to. This establishes that functional histories can create meaning conventions which possess one of the more puzzling features of normative utterances. On the other hand, none of the rules we have discussed so far are anything more than historical patterns. They have objective existence, to be sure, but there is no clear sense in which it would be wrong to violate such rules.

The critical move at this point is to directly analyze the meaning of normative statements (or the feelings they express) to see what would make them true. Simple signaling systems like the vervets' system of warning cries have correspondence maps which (extensionally) specify states of the world external to the organism, although in the case of the vervets, the signals' truth depends not simply on some "neutral" state of the world (like the presence of a leopard at some location) but, crucially, on the leopard being close enough to constitute a threat. In general, one might expect that the truth conditions for simple signaling systems tend to involve some *relationship* involving the organism(s). Moreover, there is no reason why external states need to be involved at all. The signal I experience as hunger has a correspondence rule of the sort we have been discussing, but what makes it true is just that fact that my stomach is empty. The variable states of the world outside my skin don't seem to get involved. On the other hand, if adrenalin coursing through my veins means "danger," this is true if I currently stand in some relation (being in danger) to things outside of me. According to the nonpropositional semantics of adapted control systems, all it takes for a signal to represent something is the appropriate adaptive history. Adaptation forges semantic links. Understanding the meaning of normative statements, feelings, or sentiments is a matter of understanding their function. What is it that norms actually do?

The simple answer is that they regulate. Moral intuitions regulate social behavior. Epistemic intuitions regulate reasoning and belief formation. This isn't much help because, one way or another, all representational systems regulate. What makes normative systems distinctive is that they regulate via the enforcement of rules. Which rules do they enforce? Obviously, the ones they are adapted to enforce. Now, we would seem to have meaning rules that are directed toward other rules, rather than toward world-states. The concept we need here is that of a regulatory hierarchy. Rule-enforcement mechanisms form a higher level in a regulatory hierarchy. How do such things arise?

Because of the way that nature goes about providing solutions to adaptive problems – that is, selection and random variation – the preliminary version of any system tends to be rather inefficient. But once something that is at least better than nothing is in place, optimization can commence. Small modifications of the system arise via the usual inaccuracies of biological reproduction and, barring accident and given time, those that are superior with respect to the particular function will be selected for. For our purposes, the kinds of modifications that arise fall into two categories. The first is the most familiar. The *existing* structure might be modified, for better or worse. On the other hand, instead of modifying existing structures, *new* mechanisms might arise which improve the performance of existing structures by interacting with them. The common bacterium *E. coli* has, along with an ingenious system of motorized flagella dedicated to foraging and toxicity avoidance, a collection of chemical sensors. Presumably, some of these sensors have been *added* to the existing system to improve the functioning of the older mobility system. Genes are commonly divided into two categories: *structural* genes, which code for proteins and enzymes, and *regulatory* genes, which turn them on and off. Again, there is an asymmetrical functional dependency that arises from the addition of regulatory mechanisms to a preexisting system. The particular relationship we are after here is one in which some new regulatory mechanism arises that *enforces existing function.*

Primitive Content

To pick a well-known example, in his classic article "The Evolution of Reciprocal Altruism," Robert Trivers (1971) presented a sketch of the system underlying human altruism. The proposed reconstruction of the evolutionary history, based mostly on anthropological studies of tribal peoples and laboratory studies of human moral and cooperative behavior, is as follows. The economics of cooperative behavior are such that cooperation is unstable, as exhibited in the familiar "prisoner's dilemma" of game theory.[7] Nature's initial solution has been to provide "strong positive emotions" favoring cooperation. This may be an adequate solution when one usually plays against close kin, as in Hamilton's (1964) kin selection model. However,

> Once such positive emotions have evolved to motivate altruistic behavior, the altruist is in a vulnerable position because cheaters will be selected to take advantage of the altruists' positive emotions. This in turn sets up a selection pressure for a protective mechanism. Moralistic aggression and indignation in humans was selected
>
> 1. to counteract the tendency of the altruist, in the absence of reciprocity, to continue to perform altruistic acts for his own emotional rewards;
> 2. to educate the unreciprocating individual by frightening him with immediate harm or with the future harm of no more aid; and
> 3. in extreme cases, perhaps, to select directly against the unreciprocating individual by injuring, killing, or exiling him. (Trivers 1971, 49)

Trivers goes on to suggest that a sort of "arms race" can ensue between cheating and the detection of cheating. "Sham moralistic aggression when no real cheating has occurred may nevertheless induce reparative altruism. Sham guilt may convince a wronged friend that one has reformed one's ways even when the cheating is about to resume" (Trivers 1971, 50).

Such innovative deceptions involving enforcement mechanisms create selective pressures for new detection and enforcement mechanisms, the discriminating responses of which can then be exploited, and so on. The result is a hierarchical system of controls in which patterns of enforcement are themselves enforced, resulting in many levels of regulatory hierarchy. Notice that, in principle, all this regulatory complexity can arise through the evolution of "instinctive" behaviors.

Whether Trivers has the details of the adaptive history of human altruism exactly right is tangential to the point being made here; it is well established that human social norms have some kind of hierarchical regulatory structure like the one he described and that structure is universal enough to make adaptive histories of accumulating regulatory mechanisms plausible. Our question concerns the correspondence rules of the enforcement mechanisms. What makes cheater identifications true and what, if anything, is the difference between the associated correspondence rules and those of more basic signaling systems?

Say that a mechanism of "moralistic aggression" arises and is selected to compensate for cheating (nonreciprocating) behavior. Such a mechanism is another example of an adapted signaling system, so there will be production rules governing the issuance of the cheater-recognition signal and interpretive rules specifying the appropriate response according to the design process. As before, the correspondence rules for the signals are separate from (although complementary to) the rules governing production and interpretation. The correspondence rule specifies the state in which cheater detection has been advantageous. If the enforcement mechanism has been selected specifically to eliminate the compromises to the design of the cooperative system posed by cheaters, then the cheater-identification signal is true just in case the rule governing the operation of the system of cooperation has been violated. What the signal must correspond *to* in order to be true is the *failure of a rule* of adapted design. Put another way, the regulatory system is referentially *about* the enforced rule.

Our simple formalization of biofunctional semantics can easily be extended to the enforcement of rules of adapted design. Let us say that a rule of design *fails* when one of the conditions specified by the rule is not accompanied by the indicated process. For some adapted mechanism **M**, the failure of the rule governing **M** is as follows:

$$\text{Failure of } \mathbf{R_M} : (\mathbf{condition}\ \&\ \neg\mathbf{process}) \text{ where}$$
$$<\mathbf{condition, process}> \in \mathbf{R_M}.$$

For some stabilizing mechanism (**SM**), the correspondence rule for its corrective signals $\mathbf{CS} = \{\mathbf{cs}_1, \ldots, \mathbf{cs}_n\}$ is given by

$$\mathbf{R}(\text{correspondence})_{\mathbf{SM}} = \{<\mathbf{w}\ \&\ \mathbf{m, cs}> \mid \mathbf{SM}\ \text{was selected for sending}$$
$$\mathbf{cs}\ \text{when}\ \mathbf{w}\ \&\ \mathbf{m}\},$$

where the **m** is individual states of the stabilized mechanism. (This includes the processes and some of the conditions in the general specification of the rule for **SM**.) Since **SM** was selected for stabilizing **M**, the states of the joint system **W** + **M** in which **SM** was selected for sending signals are just

Primitive Content

conditions in which some component of **M**'s rule R_M was violated. Which is to say,

$$R(\text{correspondence})_{SM} = \{< \textbf{condition} \;\&\; \neg\, \textbf{process} \text{ where } < \textbf{condition},\\ \textbf{process} > \epsilon\; \mathbf{R_M}, \text{cs} > \;|\; \textbf{SM} \text{ was selected for sending}\\ \text{cs when } (\textbf{condition} \;\&\; \neg\, \textbf{process})\}.$$

Again, the corrective signal is true when elements of the rule implicit in **M**'s history have failed.

Cheater-detection mechanisms of the sort with which Trivers was concerned provide an ideal case in point here. Systems of this sort have been widely observed in many species and, moreover, the economics driving the arms race has become fairly well understood. What this forces on us is the realization that behavioral regulatory hierarchies of this sort are ubiquitous. Consequently, the function of such systems is clear as well. This clarity allows us to build a general model for functional meaning of rule enforcement mechanisms. Two points will help forestall confusion at this juncture.

First, the rules for a trait are not a general description of how that trait has operated, but of how it has operated that resulted in positive selection. If the trait is inefficient, this may only be a description of how it has operated a small part of the time. Enforcing such a rule is just a matter of getting the trait to operate in the beneficial manner more of the time. Which is to say, rule enforcement increases efficiency in the performance of a task. Rule enforcement is optimization, and optimization when it occurs will be selected.

Second, the rules we have been suggesting are simply historico/economic patterns. There is, again, no *general* sense in which it is wrong to violate these rules of adapted design. What makes certain historical patterns of social behavior normative is that there is a higher-level signaling system dedicated to enforcing them. The trick to understanding normativity is, I have suggested, a matter of what would make a normative intuition true. When normative intuitions result from the operation of adapted rule-enforcement mechanisms, what makes them true is the failure of the rule they are designed to enforce. This might explain why we think they concern objective extracultural rules. We don't introspect this reference, of course. Rather, we observe the patterns of their responses and try to figure out what they are about. Statements of general rules are then derived from the observation of normative intuitions. (This is actually how the philosophical investigation of rules of human conduct usually proceeds!)

For those interested, the simple rule-enforcement signal provides a model with which one might begin approaching the meaning of *moral* intuitions – that cheating, lying, and murder are wrong, for instance. To be sure, morality is more than just a collection of individual social-behavior-regulating "instincts" but is at least in part a set of socially evolving sanctions of such impulses. For instance, it appears that the institution of social justice has suppressed a much older human tendency to seek retribution, although this is consistent with the claim that the *desire* for revenge has, due to its own adaptive history, conditions for proper occurrence and proper consequent behaviors as well. Which is to say, the desire for revenge may speak truly, but one might be better off or even morally bound to ignore it. Understanding morality even in biological terms will thus require understanding not only the functional meaning of the basic human social-regulating impulses but also the way in which social institutions evolve so as to regulate those impulses in their own right. Of critical importance will be the functional analysis of the system of evolving social sanctions on moral intuitions.

Our purposes here concern epistemic norms, however, so that it behooves us to turn aside from the broad architecture of rule-enforcement mechanisms in general and focus on the specific kinds of rules that epistemic norms are designed to enforce.

OBJECTIVE RULES OF REASON

Somewhere in the process of becoming disenchanted with what I think of as X-knows-that-p-epistemology, I was struck by the following interesting fact. The substantial debate over the definition of knowledge that was stimulated by Gettier's (1963) troubling paper was able to proceed not because anyone knew descriptively what knowledge was, for that was precisely the point at issue, but because philosophers shared in common a remarkably uniform sense or intuition as to what does and does not constitute knowledge. This common sense of what knowledge is allowed the testing of new definitions via counterexample. For instance, knowledge had been defined since Plato as "justified-true-belief." If, however, Jones owns a Ford and Smith believes that Jones owns a Ford because he saw Jones buy a Ford, only Jones wrecked that Ford and bought a Pontiac and sometime later was left *another* Ford by his maiden aunt – well, then, Smith has a true belief and was not unjustified in that belief. Yet, everyone knows that Smith doesn't actually *know* that Jones owns a Ford. There is something missing here. The normal thing for the trained philosophical mind to do at this point is to dig into the question.

"Ah, yes. There is something missing in Smith's alleged knowledge. What could it be? Some kind of 'appropriate' causal connection between the belief and its object? But what does 'appropriate' mean?", and so on.

For us, this is the point at which we make the critical move, to back off from engaging in the normative discussion and instead attempt to analyze the language involved. In this case, it is the phrase "knows that." What the Gettier problem highlights for us is not that we fail to have a defensible definition of "knowledge" but that the very thing that naturalistic epistemology is supposedly unable to deal with is, in fact, a very common *feeling* that people have about what does and does not constitute knowledge. For the most part, the other normative concepts of epistemology – truth, justification, comprehension – remain perennial parts of our conception of knowledge not because they are things whose existence we infer from experience but rather because they are responses to epistemic situations that anyone can be trained to recognize. Hume analyzed the perception of causality in this way. We do not perceive causation, only successions of events. Rather, the patterns of our own expectations create feelings which we attribute to external events. In Humean terms, the wrongness of irrational or confused thinking is not something we perceive as we do the shapes and movements of physical objects, but involves rather the activation of some feeling or sentiment to the effect that there is something wrong with the thinking process.

What Hume could not do – and this is a subject we will return to in the next chapter – is figure out what might make such a sentiment-based judgment *true*. We, on the other hand, are equipped with a model of primitive content in regulatory hierarchies which allows us to do exactly this – to analyze the meaning of a rule-enforcing sentiment or intuition. To flesh out the analysis of the last section for the purposes of epistemology, we need to understand not just the primitive content of the rule-enforcing mechanisms themselves but the nature of the rules they enforce as well.

Recall that in the basic scheme for the functional semantics of adapted signals, we distinguished three kinds of rules or conventions describing processes which played critical roles in the adaptive history of the signaling system. First were production rules which describe the historically adaptive conditions under which signals were produced. Although not properly part of the meaning or content of the signal, we shall see that they are epistemologically relevant, since they indicate the conditions under which the production of a signal is *justified*. Second are consumption or interpretation rules, which describe what is supposed to follow from the occurrence of the signal and which determine the intensional part of the signal's content. Third were correspondence rules, which map external conditions critical to the efficacy of the behavioral consequences specified by the signal's intension. This is the signal's extension.

These three sorts of rules, as specified, exist not only for simple signals possessing primitive content but also for any signal in an adapted system. That is to say, although we have insisted that the monolithic tracking-and-motivating signal be understood as the basic unit of biological meaning, elements of more complex systems that separate tracking from motivating and further build up indicative representations from combinations of interchangeable parts possess these three kinds of rules as well. So, for example, an indicative representation like a belief has (because of its role in some larger system) conventions governing proper conditions for its production, for what follows from it both inferentially in terms of other beliefs and practically in terms of what behaviors are likely to be efficacious in terms of satisfying needs. Similarly, elements of representations such as words and concepts themselves derive their meanings from their various roles in sentences and may individuate objects (as opposed to partitions of world-states) as the common features of the world that explain the efficacy of the various complete representations of which they are a part. The latter is Millikan's (1984) analysis of word-meanings as I understand it, and I refer the reader to her for a more complete account. The pertinent point here is that the adaptive histories of systems of representations determine rules for the production, consumption, and external conditions for the occurrence of indicative representations like beliefs, as well as for monolithic tracking-and-motivating signals like normative intuitions appear to be.

The norms of epistemology which we need to understand concern, for the most part, the production of beliefs, and it is for this reason that we need to recognize that the three sorts of rules exist for beliefs as well as for tracking-and-motivating signals. The schematic for epistemological norms falls within the general scheme for rule-enforcement mechanisms discussed previously, with the provision that the rules which the normative mechanisms exist to enforce are precisely the three kinds of rules which exist for any adapted signal. Thus, our scheme of three kinds of rules does double duty for us. On the higher level, it lets us understand first how the normative signal can both track and motivate; and, second, how it can take as its extension the failure of a rule governing the regulated subsystem. On the lower level of the regulated subsystem, the three rules provide us with three *specific* kinds of rules whose failure provides adaptive opportunity and thus subject matter for the higher level regulating subsystem.

When we confront our epistemic intuitions with particular instances of purported knowledge, we typically respond to two kinds of failures. Indeed, the identification of these two types of failures lies behind the old definition of knowledge as "justified-true-belief." We are disturbed somehow, if a belief is merely accidentally true, unjustifiably formed without the proper evidential or logical basis. Something in us responds to these kinds of cases consistently enough that everyone agrees that if a belief is unjustified, it is not knowledge. We also are certain that you can't know something that isn't even true. In both cases, the intuition that the belief does not constitute knowledge takes as its extension the failure either of the extension of the belief in question or of the rule for its production. Notice here that the primary effect of the normative intuition is not the formation of a factual belief to the effect that there is a natural property of being knowledge, and your belief does not have that property. The primary effect of the intuition is the refusal to accord a certain status to your belief, that of being known, which is critical in determining the role it can play in future processes of deliberation. Of course, we *use* subject-predicate language to express this refusal, but such opportunistic uses of language in communicating have become widely accepted in linguistics under the category of "illocutionary acts." The truth-bearing relationship derives from the function of the intuition rather than from the literal structure of the sentence with which we express it.

Implicit in the requirement that knowledge be at least justified-true-belief is the requirement that the rules for consumption be obeyed as well. We have numerous terms for failures of this sort, such as failure to understand or comprehend or grasp the meaning of a belief. So, for example, we would refuse to accord the status of knowledge to a commitment to a statement such as "porpoises are mammals," even if it were acquired on good authority and true, if one were under the mistaken impression that sharks and porpoises were one and the same. In this case, we would not say that the belief was unjustified or false, but that you somehow failed to clearly believe that porpoises are mammals, that the belief was confused, or you failed to understand the meaning of the statement.

Thus, we see a close fit between the semantic architecture of rule-enforcement mechanisms for signaling systems and the intuitive sources of epistemology's normative concerns. (Indeed, it is this conformity which makes epistemic norms simpler to analyze than moral norms.) The three kinds of rules that exist for adapted signals are precisely the kinds of rules with which epistemological intuitions are concerned. Moreover, the normative intuitions themselves have the primitive character of the monolithic tracking-and-motivating signals, albeit serving a higher-level role in enforcing the historically productive regularities of adapted signaling systems. There is some reason to think that these may be the *only* sorts of rule that epistemic norms address. Even the inference rules of logic are no more than rules for the consumption of premises and the production of conclusions, at once forming chains of rule-based deductive and inductive inferences.

If I am right, we have just sketched a way to allow a purely descriptive theory to account fully for epistemic normativity, resolving the biggest outstanding objection to the naturalization of epistemology. How was this possible? There were three critical stages. First, broadening naturalism via the inclusion of adaptive histories allowed us to incorporate a remarkably flexible correspondence theory of meaning along the lines of that developed by Millikan. Second, focus on monolithic tracking-and-motivating systems allowed us to recognize the existence and significance of primitive content, breaking the logjam created by the dogma of the word as the basic unit of meaning and of the proposition as the only kind of normatively environment-tracking content. The significance of primitive content can hardly be overstated since it may provide the key not only to the understanding of norms but also to the meaning of human emotions and the meaning of animal communications. Finally, the notion of a regulatory hierarchy invites the application of the theory of meaning to higher level rule-enforcement signals. Such signals are, from the functional point of view, obvious candidates for the neural basis of normative intuitions and they turn out to have the character we usually believe that our normative intuitions have. They are *about* objective extracultural rules for the operation of our common representational system. They have the authority of truth, and what follows from their truth is not a factual belief but a regulatory response – to return to conformity to the rule. The only way in which the character of these rule-enforcement mechanisms fails to live up to the common philosophical conception of epistemic normativity is that the rules involved are *not* eternal. They are not general laws governing abstract classes of systems. Neither are they forward-looking economic optimization rules. Instead, they are the productive historical conventions of some particular representational system which have happened to accumulate rule-enforcement mechanisms. The meanings of normative intuitions are no more than the meanings of those accumulated rule-enforcement mechanisms. As such, to understand the authority of a normative intuition like the one which compels us to reject contradictions, we need not search fruitlessly for some sort of abstract meaning-element to which the intuition is directed or some eternal rule of thought that binds any rational mind. On the contrary, we must simply consider the intuition itself, its functional design, and the conditions under which it is supposed to occur. If *this* contradiction is the kind of situation to which the intuition is supposed to object, and rejection of the contradiction is the adapted response, there is nothing more to be understood. The normative element of knowledge has been accounted for in the design (as opposed to the actual functioning) of the rule enforcement mechanisms governing human knowledge.

FUNCTIONAL FAILURE AND PRIMITIVE CONTENT

The most general sorts of objections to the analysis of epistemic norms given here have to do with the extent to which our epistemic behavior is learned rather than innate and whether there is reason to think that *in principle* the essence of normativity is beyond the scope of factual inquiry. Both of these concerns are widespread, and making the case for the relevance of evolution to meaning and of naturalism to normativity requires careful consideration. But before turning to these questions in the next chapter, the point needs to be made that the functional analysis of norms can be grounded in other ways than on selective histories.

The central mystery about norms or ought-statements has been what would make them true, given that they motivate directly. The approach taken here is to begin the analysis of meaning with signaling systems involving simple tracking-and-motivating signals. I have also, for reasons which should be apparent from the models in Part II, chosen to ground my account of the most basic norms in adaptive biological histories, since I see no way for any element of a representational system to be reliable with respect to coping with the "world out there" other than to have been selected through interaction with it, if only vicariously. Yet, while naturalism compels one toward such a conclusion, not everyone is that kind of naturalist. For those others, there is still something here of interest.

What one needs to understand norms via primitive content is some way of assigning rules for the production, consumption, and external conditions for signals in simple systems. What we have found is that one cannot adequately assign such meaning with the resources given by twentieth-century logic, which has taken the word as the basic unit of meaning and the proposition as the only truth-bearing whole. On the other hand, *any* account of the purpose of a sender-receiver signaling system as an environment coordinating system can provide fallible standards which can be used to specify primitive content for signals. Presumably, if you think normative intuitions have a purpose, you think that purpose is to enforce some rule *by* some particular kind of regulative response. So, if you can assign a fallible purpose or function to normative intuitions, then you can specify primitive content for those intuitions which specifies both extension in terms of the failure of some rule which they are supposed to enforce and intension as the direct regulating response.

Rather than being a naturalist, one might, for instance, be a theist who thinks that our normative intuitions were designed by our maker to get us to adhere to certain rules. The rules governing the meaning of intuitions exist in the intentions of the diety, but the content can be understood as primitive. In such a case, it is not that the intuition expresses the proposition *that* God intended us to adhere to the rule, from which we infer that we ought to do so. Rather, the content of the intuition simply specifies conditions under which it is supposed to occur and the response that is supposed to follow. No inferential process need mediate.

Alternatively, a rationalist might suppose that an agent deliberately or tacitly participates in a system of social controls which involves the operation of conscience. What do the voices of conscience mean? They seem to command the respect for property, but they also pose as though the commands were true, as though they had a kind of objective authority beyond their power to compel. If one knows the purpose of the system in which these voices play a regulative role, then one can derive from that purpose rules for the production, consumption, and correspondence of those voices. Satisfaction of the correspondence rules constitutes truth and the consumption rules determine what follows: when the voice speaks truly, what it says is "desist!"

The essential point is that what I have called the critical move – to step back and ask what would make the intuition true in the context of a tracking-and-motivating system – is not a move that is exclusively available to the evolutionary naturalist. To be sure, evolution offers a particularly well-grounded notion of purpose, function, or convention, but it is having a notion of purpose that matters for this analysis, not where you get it. What the consideration of our status as thinking *animals* has done is offer the tracking-and-motivating signal as a candidate for the basic unit of meaning, especially since animal communication has predominated this form. This is not an insight that one is likely to have via introspection, although if one looks one can find considerable leanings of this sort (e.g.) in Wittgenstein (1953). Having gotten the idea, from whatever source, that intension and extension derive from a signal's role in the larger system, one is then apt to realize that what I am calling "primitive content" – the conjunction of extension with direct behavioral or inhibiting intension – is the simplest kind of content, and that the existence of this kind of meaning pretty much dissolves the mystery surrounding the meaning of normative intuitions and utterances. Propositional representation is not the only kind of representation that goes on in peoples' minds, and it is high time we realize this.

Such bold statements would seem to require far more defense than I can possibly give in one remaining chapter. My intent in stating things so adamantly is not, as it might seem, to attempt to convince merely by strong statements, for I respect my reader too much to think that that is possible. My intent, rather, is to attempt to convey my growing conviction that we have gotten badly off on a rather anthropocentric track in our understanding of meaning with the emphasis on combinatorial syntax, on the word as the basic unit of meaning. To make things worse, such a paradigm for meaning fails to accommodate much that is distinctively human, critical to our conceptions of ourselves. Our minds are not unified as the old conceptions of detachable souls and ideally rational agents would have it. Rather, we are complex accumulations of partly integrated systems of varying antiquity, more junkyard than axiomatic system. Moreover, there are parts of us that are at once distinctively human and yet cry with the primitive voice of animal warning. We reject our animal nature at the risk of misunderstanding what we take to be our most essential humanity.

9

Is and Ought

The task for the previous chapter was merely to *articulate the hypothesis* that there is a basic kind of meaning – primitive content – and that the normative intuitions regarding standards of rational thought and epistemic justification which constitute the primary objection to epistemological naturalism, in fact, possess this kind of content or meaning-rules, rather than the more elaborate propositional content in terms of which we have been attempting, unsuccessfully, to understand them. In the simplest terms, animal warning cries and human feelings are not merely brute expressions of distress but possess conventional, as opposed to natural, meaning.[1] Try to understand normative intuitions as a kind of warning cry, sent by adapted rule-enforcement mechanisms, and you have the essence of the proposal. We saw that the three sorts of conventions implicit in the functional history of any adapted signaling system map nicely onto the truth, justification, and comprehension of beliefs. What Chapter 8 did not attempt was to establish that our normative intuitions, in fact, express this kind of primitive content and that the content they possess largely conforms to our usual sense of such things. Indeed, as an empirical hypothesis, its confirmation falls far beyond the scope of these short chapters, depending as it does on obscure facts of adaptive history, neurological architecture, and the study of human emotions. Preliminary to any such investment of time and resources, one must at least answer theoretical objections to the effect that this *cannot* be correct.

This is the general opinion among philosophers – that naturalistic theories of normative standards, like that sketched in the last chapter, cannot possibly be correct. The general support for this opinion is twofold. First is the fact that previous such theories like Hedonism or Social Darwinism have tended to say that certain things are right which seem to us very clearly to be very wrong – they quickly seem to run afoul of standard counterexamples. That no one has managed to get it right yet is not particularly surprising because what I believe to be the correct answer depends so critically on the integration of historical relations into our empirical worldview, an integration which is still quite in process. Second, there are several philosophical arguments which purport to prove that such account cannot, in principle, be correct (or, at least, cannot be complete). These arguments are usually credited to Hume and Moore, although they often seem to have taken on a life of their own. In addition, the extreme flexibility of human behavior continues to cast doubt on the significance of evolution to the understanding of what are taken to be the more refined aspects of human cognition such as moral and epistemological judgments.

In this chapter, I try to explain why the philosophical and empirical arguments do not work and, unfortunately, much of that explanation must occur in the context of moral theory in which the arguments arose. In a nutshell, Moore's arguments were always plagued by assumptions of phenomenological transparency; related worries regarding the inability of science to address "what it's like" to be human tend to misunderstand the nature of theories in general, expecting something more than the mere *representation* of the world. Hume's analyses, while cogent in the context of eighteenth-century empiricism, fall apart in the face of the historical relations that have come to the fore in modern biology. Selective histories can supply standards for primitive content in at least two ways, despite (or even because of) the flexibility of human behavior. The particular criticisms addressed in this chapter were chosen on the basis of how often they seem to come up, the extent to which they affect naturalistic theories of the normative in general (as opposed to cutting particularly against naturalized ethics), and whether they highlight certain important positive features of the approach advocated here. I apologize in advance to those readers who find their favorites unrepresented. Beyond merely defending evolutionary naturalism against various objections, I argue (1) that the only in-principle limitation on naturalism is not being able to *participate in* normative discourse directly, (2) that to take this as a failure of naturalism constitutes a misunderstanding of the function of theoretical representation, and (3) that evolutionary naturalism is able to actually explain the gap between *is* and *ought,* providing several other useful analyses as well. For those interested, the way in which functional histories define categories that *supervene* on natural categories is explained. I close with a discussion of how a general or overriding normativity could exist which could adjudicate conflicts between normative subsystems.

MOORE'S "OPEN QUESTION"

For many philosophers, the reason the primitive content hypothesis cannot be right is not due to any particular feature of the construction itself. Instead, the philosophers I am thinking of bring their doubts to the table with them, previously convinced that scientific facts can do little to illuminate the basic normative questions. The reasons they often give involve reference to the famous "open question argument" of Moore, although, to be sure, the argument often seems to be code for what is actually a cluster of related worries, the clustering effect providing the argumentative weight that the individual arguments cannot.[2]

Moore, who spent most of the first half of the twentieth century doing philosophy at Cambridge, recanted the Bradleyan idealism of his youth to become the architect of commonsense realism and ethical intuitionism. His arguments concerning the naturalization of norms come not from his epistemological realism, but from his ethical theory. Perhaps most influential has been his description of the so-called naturalistic fallacy. Moore's intuitionism insisted that goodness was a nonnatural, unanalyzable property that could not be reduced to any sort of physical properties and was apprehended directly through ethical intuitions. Hedonists and utilitarians, on the other hand, identified the good with pleasure. Moore accused them of committing the naturalistic fallacy, of moving from the *attribution* of goodness to pleasure to the *identification* of goodness of pleasure. Critics of epistemological naturalism infer from the broadness of Moore's argument that it applies to epistemic "goodness" as well.

Discussions of Moore on the naturalistic fallacy typically involve a certain amount of wondering just exactly what the fallacy was supposed to be. On the surface, it might seem as though moving from attribution to identification is just the common logical fallacy of affirming the consequent – of moving from the claim that if something is pleasurable, then it is good to the conclusion that whatever is good is pleasurable as well. Even apart from the absurdity of thinking that all of utilitarianism rests on such a simple logical error, there is reason to think that the fallacy is of a different sort. Indeed, Moore's overall stance seems to indicate that *any* naturalistic account of the good must be mistaken, which would be the case if Moore's intuitionistic principles were correct. Establishing those principles requires a separate argument.

The separate argument is the "open-question argument," which is directed not only at the alleged utilitarian identification of pleasure with the good, but also toward any possible naturalistic account of the good. Unfortunately, things are no clearer here than they were with the naturalistic fallacy itself. Ethicists Darwall, Gibbard, and Railton (1992) note that

> ... it has been known for the last fifty years that Moore discovered no *fallacy* at all. Moreover, Moore's accident prone deployment of his famous "open-question argument" in defending his claims made appeal to a now defunct intuitionistic Platonism, and involved assumptions about the transparency of concepts and obviousness of analytic truth that were seen (eventually by Moore himself) to lead inescapably to the "paradox of analysis." (Darwall et al. 1992, 115)

Nonetheless, they urge that despite the failure of the argument itself, one has to take it seriously; applied on a case-by-case basis, it still compels. By now, I assume you actually want to know how the argument goes.

Roughly, the argument is that whether one claims that the good is pleasure, or the greatest satisfaction of revealed preferences for the greatest number, or the stability of the state, or the survival of the species or the biosphere, or the progress of evolution, one can still ask "but is it good?" If one can still ask "is it good?," then apparently, we haven't gotten at the question of what it *means* for something to be good. The argument as stated assumes that since we know what good means, we should recognize any true general characterization of the good. Such an expectation depends on just the unfashionable Platonism and assumption of the transparency of concepts referred to by Darwall et al. So why is the argument still worthy of consideration?

The answer seems to be that the argument gets at something that actually *is* important. In a nutshell, the question of goodness is, phenomenologically, a separate question from the question of whether it is pleasurable, promotes stability, or what have you, and this seems to be somehow important in understanding what goodness is. The conviction that an act stabilizes society (or whatever your account claims is the basis for the good) is not the same conviction that the act is good. Consequently, any such account is incomplete, misses the point, fails to get at what it *means* for something to be good. We find consistently that when we are presented with mere naturalistic *descriptions* of normative human processes, we are never "engaged" by the descriptions in the way that we are engaged by the norms described. Even if I had all the necessary facts to describe the evolutionary history of some neural circuit whose activation is expressed in your moral outrage at seeing a child tortured, still you might justifiably feel that my description has missed something essential. My description of these processes simply does not affect you in the way the normative intuition itself does. It seems, therefore, that there is something inherently missing in descriptive accounts of normativity – namely, the normativity. So while the reasoning that Moore employed and the conclusions he drew don't fly for us these days, the procedure he suggested continues to compel assent to, at least, something like the conclusion he wanted.

Consequently, addressing the open-question argument requires more than simply refuting its surface form, at least if one is interested in getting at what is really bothering people rather than just sending them back to the drawing board. The refutations are fairly simple. As noted earlier, without the assumption of introspective transparency, the argument never goes anywhere. Why think that you are somehow automatically able to recognize the true account of ought-statements? If their meaning is determined by external standards, then you can be mistaken about them and thus fail to recognize them when they are presented to you. Another refutation consists of the related observation that if such a requirement is allowed as telling, then no scientific account of any mental phenomenon can get it all. (There are many people who hold this opinion. I shall address their argument presently.)

Finally, the particular primitive-content hypothesis allows a specific and direct refutation. The hypothesis is a hypothesis about meaning. Theories of meaning have a particular immunity to the open-question argument; to accept such a theory *just is* to accept an account of the meaning of "is wrong." How, then, can one claim that we have not said what it means for something to be wrong? Indeed, we shall see shortly that the primitive-content hypothesis actually allows us to explain exactly *how* naturalistic descriptions fail to fully capture the meaning of normative intuitions, although without leaving anything out. We will see later, from within evolutionary naturalism, the real truth behind the open-question argument. Before getting to that, we need to clear up some other confusions that haunt the naturalistic analysis of the normative.

THE PHENOMENOLOGY OF AGENCY

For some time, I was convinced that the real reason people keep bringing up the open-question argument was because it points out that the scientific descriptions never seem to fully get at "what it's like" to be a moral or epistemic agent (descriptions of normativity leave out the normativity). The sense in which we feel "bound" or "engaged" by normative rules seems wholly different from the effect of the dry tone of scientific descriptions. The passion, if you will, of the intuition is necessary to the full understanding of the intuition. Science, dedicated to the dispassionate description of value-neutral fact, cannot possibly capture the normative without it. This argument, which I infer more from argumentative stance than from any explicitly defended argument, is virtually identical to arguments in the philosophy of mind and cognitive science which purport to argue that science can never fully account for the quality of mental experience.

The classic argument is a thought-experiment due to Frank Jackson (1986), and it goes something like this. Mary is a twenty-third-century cognitive scientist, and Mary knows absolutely everything there is to know about the brain, its neural structure and chemistry, its physical changes in development and learning, and how it interacts with perception and behavior. Make this knowledge as detailed as you would like, down to the subatomic "particles" if you think it will help. The problem with Mary is that she has lived all of her life in a small room in which everything is black and white, or shades of grey. By hypothesis, Mary knows everything there is to know about what it is to see the color red, what wavelengths are involved, what neural responses are involved, but never having herself seen the color red, Mary does not *know what it is like* to see the color red. Mary's exhaustive scientific knowledge leaves out some essential kind of knowledge – the knowledge of what it is like. Q. E. D., no scientific theory can ever explain everything about mental experience. This is called "The Mary Problem."

My favorite response is one I saw given by noted "neurophilosopher" Pat Churchland. Seems that Pat had bought a new microwave oven which came with a microwave cookbook. In the front of the cookbook was a little essay by Betty Crocker explaining how the oven works. Betty explains how the element in the oven gives off electromagnetic waves carefully calculated to be at the exact wavelength that excites the water molecules and gets them moving around. So far, so good. Betty goes on to explain how all that movement causes the water molecules to rub up against each other, and all this rubbing causes friction, and friction, as everybody knows, causes heat. The heat cooks the food, and so on. The problem with this is that Betty had apparently slept through thermodynamics, or else had forgotten that heat just is mean kinetic energy. At the point where the molecules were moving, she was already done explaining where the heat came from and didn't even know it, and that's the moral of the story. Sometimes you really are done explaining things and you don't know it. It may not seem like it to you, but heat just is mean kinetic energy. Let's call this "The Betty Crocker Problem."

Churchland's point in bringing up this rather embarrassing episode in the literary career of poor Betty was to make the same point with respect to neuroscientific explanation. Once you have a complete explanation of all the neural processes, of the sort Mary was supposed to have, you are done. You might think that there is something more, some further stage of explanation which will bring things closer to something you can relate to, like the familiar heat of friction that Betty needed for closure. Science isn't like that. Scientific explanations for familiar phenomena are frequently unfamiliar. Heat is not hot, but just motion. Solid objects are not really solid. Physics does this to us all the time. Why expect neuroscience or, in our case, naturalistic epistemology, to be any different? Why expect the correct description of a phenomenon to be obvious in its correctness?

The response from people who think that the Mary argument and the open-question argument are really onto something should be to say that Churchland's point merely concedes what they were arguing in the first place – that there are limits to scientific explanation and those limits leave out the mental or the normative (or both). The naturalist and the phenomenologist face each other across the divide between dry material fact and the true essence of what it is to *be* a human being. Nothing here has shown this gap to be illusory.

The problem with such a response is not that it insists that the arguments indicate a gap that may be unbridgeable. On the contrary, I would agree that there is such a gap. The problem is that the nature of the gap is left mysterious, and the significance of the gap is taken to be a criticism of naturalist materialism. Let's address the latter problem first.

Consider: every description or explanation leaves out something important, namely, the very thing that is explained or described. This constitutes no failure or shortcoming in the description. On the contrary, it is of the very nature of representation in general to be different from the thing it is about. Explanations of rocks leave out the rocks. Explanations of human qualitative experience leave out the experience. Explanations of norms leave out the very substance of normativity. This is all as it must be. Why does it ever seem like a criticism of science?

The answer, I think, is twofold. First, there is something else to "know" about seeing red, something that Mary does not know, but this is not descriptive knowledge at all. Rather, it is a kind of knowledge by acquaintance. To know what it is like to see red, or to be a bat,[3] is to have been that thing or something similar. The knowledge involved is not a matter of having a description or a theory, but of having a memory derived from one's own experience. To have "been there" in this way is not something that *any* theory can give. What it takes, instead, is a manipulation. As a consequence, the demand that science tell you "what it's like" is unreasonable in the extreme, since it is unreasonable to expect it of any pure representation.

Second, representing representational systems is a curious business, especially when the system we are representing is our own. The reason the failure of a theory of granite to capture the granite itself does not seem troubling is that there aren't any options for human "being there" on the other side of the representational gap. The relationship between granite and picture-of-granite is just the classical straightforward relationship between any representation and its object. This is what I meant by a "pure" representation. On the other hand, when we begin to describe elements of human experience there is something on the other side of the gap that represents a possibility for us. The picture-object relationship is complicated when the object itself is a picture.

The real crux of the matter is not whether science can, in principle, account for mental life; that is, whether the mental is just one part of the material. The real issue concerns what one wants to do with the theories one builds. When one is concerned with a system of representations like human normative intuitions, utterances, and discussion, one may do two sorts of things. One might attempt to build a powerful representation *of* the intuitions and discussion, or one might attempt to build representations that *participate in* the system in question. In philosophy, this just boils down to the difference between descriptive and normative theory; in psychology, between intentional psychology and cognitive science. What the Mary argument and the phenomenological interpretation of the open-question argument show, as it bears on the project of this book, is simply that naturalistic accounts cannot *participate in* normative systems in the way that normative ethics and epistemology try to. As far as these arguments go, however, that is all that cannot be done, that is the only principled limit to naturalism that follows. All that these arguments amount to is the observation that the naturalist is not doing normative theory, that he is simply describing rather than attempting to participate in the systems concerned directly.

What the naturalist can do that the normative theorist has been unable to do is actually to explain the nature of the gap. The distinction between description of representational systems and participation in them is the core of the solution. The specifics of how human representational systems work often makes this hard to see.

THE NATURE OF THE IS-OUGHT GAP

Begin by keeping in mind the notion of primitive content. The basic adapted conventional coordinating signal both tracks and motivates, and its history determines fallible rules for tracking and motivating. Paradigmatic human decision making breaks up this functionality into multiple parts. First, tracking or indication is separated from motivation, as evidenced in the belief-desire distinction. Second, these indicating representations possess combinatorial syntax, words and concepts, subjects and predicates, and attribute properties to objects. Scientific description elaborates the indicating function alone, explicitly foregoing any direct connection to action. This indicating system has become powerful and precise, capable of the fine-grained individuation of world-states, properties, and object-types.

Suppose we direct this indicating system toward another representational system exhibiting primitive content, such as the vervet's warning cries. The vervet's signaling system, although effective, is crude in the partitions it imposes on the world. The world is in one of four states: leopard, eagle, snake, or none of the above (silence). Scientific language is flexible enough for us to be able to create representations which share the same extension as any one of the warning cries. We can fine-tune this overlap. Does the cry share its extension with our sentence "there is a leopard here," with "there is a leopard within 300 yards," with "there is a hungry leopard within 200 yards," or some other? Answering this question is difficult because it depends on the historically stabilizing function of the warning system, and we have to infer this from current behavior. Whatever the case, it is undeniably the same world to which we and the vervets are referring, and scientific language is flexible and precise enough to duplicate the extension of the vervet's cry pretty closely if we know enough about the function of the system.

What scientific language cannot do is, in the same signal, duplicate the intension of the vervet's cry. The descriptive sentence which is coextensive with the vervet's cry has a very different intension, and thus a different kind of content altogether. Our sentence has all kinds of implications regarding facts about leopards that are entirely absent from the vervet's cry. The vervet's cry directly means the evasive action itself, which our sentence would not imply, at least with the same immediacy. Nonetheless, we can use the extension of our powerful language to pick out the intension of the vervet's signal. We say that the signal means-extensionally something like "leopard here now!" and means-intensionally "run up a tree." We can also note that the intension and extension are combined into a monolithic tracking-and-motivating signal in the way they are due to some selective history. This would seem to be an exhaustive specification of the meaning of the signal – subject, of course, to the usual uncertainty about the facts, but there is no theoretical problem here. True, our inability to construct within the rules of our language a single signal with the same content as the vervet's cry means that we can't really communicate with them. We can't participate in their system with our scientific description. On the other hand, our descriptive understanding of the vervet's signal tells us enough about the conventions governing their system to participate in it using their language. However useless our signals may be for participating in their system, they certainly can tell us *how* to participate in their system and get it right. Supposing, of course, that they would let us.

Theoretically, descriptions of normative systems are no different. Subject to our access to the relevant historical facts which determine the meaning conventions, we can use our descriptive language in the same way as with the vervets. First, descriptions of the conditions under which rule-enforcing signals are supposed to be sent allow us to create descriptive sentences which are coextensive with the intuitions. Once again, as tokens in a system with a different function, we cannot in the same sentence duplicate the intension of the intuition. (This is why the open-question argument is compelling.) What we can do is use the extensional specificity of our language to describe the intensional consequences of the intuition. (Notice that individuals in communicating with others may frequently use descriptive language in the same way, expressing normative intuitions in the form of commands which describe the appropriate response.) Finally, we observe that intension and extension are combined directly to form primitive content, unique to the function of the particular rule-enforcement mechanism. What has been left out of the explanation? Nothing. To be sure, our descriptions cannot be inserted into the normative system itself, despite the fact that they fully specify (if not "capture") the meaning involved. Our descriptive sentences have a different sort of content because of their own function, but their extensional power allows full specification of various primitive contents, yet not the ability to *translate* from the "languages" that express them.

Here is the tricky part. This is where I said with respect to the vervets' system that our account told us enough to participate in their system correctly. If the parallel is as complete as I have indicated, the same should be true of the human normative system. Does this mean that you can get an "ought" from an "is"?

No. Presumably, the contents of a normative statement derive from the intuitions they express. Since meaning is conventional and conventions are historical, these intuitions derive their content from their functional history. No matter how much I understand about that history, no matter how completely I can describe the unique conjunction of intension and extension that forms that content, my descriptions do not share history with the normative intuitions in the right way to share the conventions required for content. Just as my understanding of the vervet system moves me no closer to being a vervet, understanding normative systems moves me no closer to possessing them. This is, again, just the point that descriptions of systems are not to be confused with the systems described. But, given that I *already* possess such systems, the descriptive account allows me to understand more fully the meaning of the intuitions themselves. We ordinarily issue them and respond to them without knowing what makes them true. The descriptive understanding adds dimension to the already existing pattern of responses. It will not make me a good person. It will not convince me to act rationally if I am not already committed to doing so. But it may, if nothing else, restore my confidence in my preverbal conviction that there are rules for being a human being, for correct reasoning and social behavior, and that these are not pure inventions of culture but derive from my place in a larger whole. These may not be the eternal standards that many have sought, but for me they will do.

HUME'S ARGUMENT BY ELIMINATION

More formidable than Moore's open-question argument are supposed to be the arguments of David Hume to the effect that empiricism can find no deeper basis for moral (and, by consequence, epistemological) judgments than mere sentiment. Given that Hume is one of the heroes of this book, this might seem to be something of an embarrassment. Not so. Hume's argument was quite telling, but only according to the knowledge of his time, a hundred years before Darwin. Moreover, his very stance indicates to me that he might welcome the account offered here.

The classic statement of the argument occupies Book III, Part I, Section I, of *A Treatise of Human Understanding,* Hume's first published work. It goes as follows: All perceptions (occurrences in the mind) are either *impressions* (the immediate and involuntary proceeds of sensation) or *ideas* (the possibly transformed residue of those impressions). Necessarily, moral relations must either exist (1) between ideas (the domain of Reason), (2) between the objects in the world that cause impressions, or (3) between some state of the mind (idea) and some object in the world. Hume argues that (1) and (2) are not plausible grounds for moral judgment, for in the case of the former, it is clear that while moral judgments have direct consequences for action, the judgments that Reason makes between ideas do not. For the latter, if relations between objects were sufficient to ground moral judgment, then (for instance) incest in animals would be as blameworthy as it is in humans. It is no use to point out that animals lack the Reason to perceive the "turpitude" of the act, for this presupposes that there is some extrarational turpitude there to be perceived in the first place. By elimination, this leaves only (3), and from this Hume derives two conditions on any moral system.

> *First,* As moral good and evil belong only to the actions of the mind, and are deriv'd from our situation with regard to external objects, the relations, from

Is and Ought

> which these moral distinctions arise, must lie only betwixt internal actions, and external objects, and must not be applicable either to internal actions, compared among themselves, or to external objects, when placed in opposition to other external objects. ... Now it seems difficult to imagine, that any relation can be discover'd betwixt our passions, volitions, and actions, compared to external objects, which relations might not belong either to these passions and volitions, or to these external objects, compar'd among *themselves*. ... But it will be still more difficult to fulfil the *second* condition, requisite to justify this system. ... According to the principles of those who maintain an abstract rational difference betwixt moral good and evil, and a natural fitness and unfitness of things, 'tis not only suppos'd, that these relations, being eternal and immutable, are the same, when consider'd by every rational creature, but their *effects* are also supposed to be necessarily the same. ... We must also point out the connexion betwixt the relations and the will; and must prove that this connexion is so necessary, that in every well-disposed mind, it must take place and have its influence; tho' the difference betwixt these minds be in other respects immense and infinite. (1978, 465)

Satisfying the first condition requires that we be able to specify some *unique* relation that holds between moral volitions and the external states which give them moral sanction. The second requires that we be able to make sense of *law-likeness* of those relations, despite the vast variation observed in human behavior.

Now, Hume's primary target was rationalism with respect to morality, and it is ironic that his argument has done so much to motivate rationalist *resistance* to the naturalization of our understanding of morality. This focus on rationalism, combined with his own account of natural laws as invariable regularities of experience, is responsible for his emphasis on the *necessity* and eternality of the moral relations in the second condition. As such, directing Hume's argument against a naturalist conventionalist theory of normative standards is a bit odd. For the naturalist, the parallel difficulty arises when making sense of the moral rule *applying* in every case. Invariability is one, but not the only, way to get this.

Hume's analysis assumes, along with virtually all naturalistic analyses of normative structures, that naturalism can only consider *occurrent* relationships, since it is only occurrent relationships that are causally efficacious. This assumption dates back to the birth of Western science in the seventeenth century. Explanation of the world was to consist of the occurrent mechanical interaction of material substances. Both gravity and electromagnetism proved initial embarrassments to this stricture. What Darwin's theory of the adaptation of phenotypes and the more recent functional analysis of evolutionary *design* brings to the fore is the indispensability of adaptive histories in understanding human nature. Hume had to consider history irrelevant to empirical understanding just as did his contemporaries. History has taken on a new legitimacy and significance in the years since Darwin.

The first difficulty, that with individuating the moral relations, is solved via the consideration of *homology* – similarity due to common descent – and via the primitive semantic relations provided by adaptation of signaling systems. The panda's thumb is not a thumb, although it may look like one and work like one, because it is not a homologue of true thumbs. "Thumb" defines a natural category via homology. The reason that the rules of morality and reason apply only to humans, and not to rats or oak trees, is that the rules apply to a homologous family of regulatory faculties which only humans carry. Despite any occurrent similarities between the behavior of humans and the behaviors of plants and animals, the rules of morality apply only to the former and not to the latter because only the former have the appropriate history. Only humans are designed to be governed by the rules of morality or, rather, it is only human evolutionary history in which the governing rules of morality have emerged via the accretion of controls on behavior, and the function of moral intuitions has been to regulate interactions between members of the same species. (The point is the same even if human culture plays a large role in sanctioning controls on social behavior.) Consequently, it is no longer "hard to see" how there might be a relation that exists between ideas and states that does not exist between states or between ideas. Such relationships, whose former obscurity sounded the death knell for naturalistic ethics, is the relationship of central importance for functional semantics as well as evolutionary taxonomy.

The second difficulty, that with accounting for how the rules apply even when they are not followed, is solved in a similar manner. The applicability of the rules of adaptive design depends not on *occurrent* adherence to them (as many have presumed) but on historical performance and the accumulation of enforcement mechanisms. Moreover, for the rules to apply, it need never have been the case that they were *always* followed. Conventions do not arise via the invariability of their governance, but via the contributions of their historical governance, however partial, to the adaptive contributions of the faculties to which they pertain. Similarly, the semantics of moral utterances do not depend on those utterances always having been heeded, or even always having been true. They depend on the nature of the contributions they have made to the functioning of the moral system, however unreliable or unheeded they may have been. Again, this involves the distinctive significance of functional histories like those that drive natural selection. Formerly, there were only two ways to understand rules within naturalism: rules were either stipulated or observed according to their invariability. Adaptive histories provide a third source of rules, which allow for variable adherence and need never have been stipulated.

It appears, then, that in both cases the apparent difficulty in seeing how science can account for objective norms lies in failures to consider the importance of evolutionary history in the design of human behavioral regulatory systems. Hume, however, can hardly be blamed for this oversight, coming as he did prior to the rise in popularity of evolutionary theories in the nineteenth century. Indeed, much as Hume's "pre-established harmony between the course of nature and the succession of our ideas" begs an evolutionary account, so does his naturalistic account of morality as the operation of an autonomous moral sense. Especially in his later work, it is clear that Hume did not intend to propose a subjectivist moral theory, but merely to point out the failings of rationalism and attempt to promote a naturalistic replacement. One imagines that he would find the preliminary account proffered here congenial, if not compelling.

GENETIC AND CULTURAL DETERMINISM

One of the more common objections to any evolutionary account of human behavior consists in the claim that human beings have somehow broken free of the dictates of their genes and, thus, of their evolutionary histories. This is supposed to make evolutionary accounts of human behavior, not to mention normativity, irrelevant. My answer to this objection has two parts. First, we have not broken free of our evolved nature, it is just that that nature is flexible. We are able to create and diverge from past patterns *because* of our genetic endowment, rather than in spite of it. Second, selection is not purely a process of genes but can happen on the cultural level as well. To be sure, this brings with it the dangers of runaway processes of the sort which inspire meme theorists. Nonetheless, cultural selective histories of signaling behavior involve stabilizing functions and thus provide the necessary semantic conventions for primitive content just as surely as genetic histories or as any other specification of fallible function.

Clearly, one of the most remarkable things about human behavior is how variable it is under cultural influence. How does one go about integrating this remarkable fact into evolutionary stories which seem to proceed as though all behavior were instinctive? The answer is actually quite simple. The systems that regulate human behavior are designed to be flexible. They are designed to be able to accommodate environmental novelties. They are designed to allow the transmission of information between conspecifics via language. They are designed to allow the formation and adoption of rules of social behavior. And they are designed to provide for the enforcement of rules so formulated and adopted. Doesn't this flexibility threaten to break loose the systems involved from their adaptive histories and, thus, from the correspondence rules those histories provide? Not necessarily.

Consider that most arbitrary of social rules, the traffic convention. In the United States, we drive on the right. In England, they drive on the left. It seems there is something wrong with not following those conventions when you are in those places. Two questions arise: what exactly is it that is wrong with driving on the wrong side of the street? And aren't these standards just the arbitrary dictates of culture?

Intuitively, there seem to be at least three reasons why driving on the wrong side of the street is wrong. First, it's stupid. Second, you pose a danger to others. Third, there is a convention (ensconced in law) which *says* that it's wrong to drive on that side. In all three cases, it is possible – given the appropriate adaptive history – for the judgment to be semantically grounded in that history. In the first case, if humans are equipped with a normative system which rides herd on flexible instrumental behavior and corrects it in cases where the function of instrumental behavior is compromised, and if "is stupid" is just a linguistic proxy for the correcting signal in that normative system, then it *is* stupid to drive on the wrong side of the street just in case the appropriate part of the function of instrumental behavior is violated. In the second case, if humans are, in fact, equipped with a normative system with the function of minimizing the danger we pose to others, and if "is wrong" is a linguistic proxy for the correcting signal in that system, then it is wrong to drive on the wrong side of the street just in case we are posing a danger to others. Finally, if humans are equipped with the abilities to formulate and follow conventions, and there is a normative system in place whose function is to enforce conventions so adopted, and "is wrong" is a linguistic proxy for the correcting signal in that system, then it is wrong to violate the convention just in case the system of convention following that the normative signal is designed to regulate is, in fact, not functioning according to design. If all three of these hypotheses seem plausible, then perhaps it is objectively wrong to drive on the wrong side of the street in three *senses* (according to three semantic mappings). The nice thing about traffic conventions (and one of the reasons they are so stable) is that the three sets of norms seem to agree. This is not always the case, however.

As for the second question, which side we drive on is *of course* an arbitrary dictate of culture, but it is not *just* that. It is an arbitrary dictate of culture that

Is and Ought

plays a small but decisive role in governing the behavior of an immensely complicated system of behavioral controls. It is an arbitrary dictate of culture that may do a good or bad job of regulating that system from the point of view of the system's design. And *if* there is a normative signaling system in place whose function is to evaluate arbitrary dictates of culture vis-à-vis their efficacy in contributing to the function of the systems they regulate, then accordingly, some rules are objectively, truly bad, and others good, at least in the sense determined by that normative enforcement system. Once again, however, rules of adapted designs are not normative in general. The proposal is that true attributions of wrongness are the expression of normative systems whose function it is to enforce those designs.

Thus, human flexibility under culture is not a matter of us being free of the dictates of our evolutionary history, but of that history providing us with the means to be flexible. Where there are adaptive histories, there are rules that apply, and the very fact of functioning flexibility implies regulatory hierarchy. On the other hand, selection on the genetically heritable basis of human nature is not the only source of fallible conventions for primitive content. Indeed the very notion of a stabilizing convention was developed first in the context of what we have been calling cultural selection.

As I was at pains to emphasize at the end of the last chapter, *any* criterion for fallible functional standards can determine primitive content. This is the second part of my answer to the "cultural flexibility" objection. Natural selection is in some ways the most theoretically satisfying such source, since it tends to increase the reliability and exploitability of the tracking relationship with respect to the material well-being or stability of the system. Nonetheless, natural selection in the cultural sphere, which was discussed at length earlier in this book, also results in selective histories which can determine meaning conventions. Consider traffic conventions once again. Flashing yellow lights mean-extensionally that some hazard exists and mean-intensionally "proceed with caution." It is possible, of course, that some individual at some point designed such signals to have this meaning, but it is also possible that no such individual ever existed, or that usage has diverged from the original intention. We can still figure out the extension and intension of the signal simply by observing its selected function. Once again, the function is not just anything that the signal does, but what it does that accounts for the continued existence of flashing yellow lights: the existence of hazards, the need for caution in the presence of those hazards, the reliable indication of hazards by flashing yellow lights, and the customary response to those lights of proceeding with caution. The conjunction of these factors explains the persistence of flashing yellow lights and, consequently, the primitive content of the signal can be derived from the historical function that explains that persistence. In this way, cultural evolution is just as powerful as biological evolution in providing alternative kinds of (primitive) content which can help us understand the meaning of norms.

For instance, as part of their training, scientists learn to obey a variety of restrictions on the sorts of claims they make. Most of these restrictions cannot be exactly formulated and, indeed, seem to evolve along with the discipline itself. As a discipline becomes better established and furnished with standard methods and background theories, standards of evidential support and statistical rigor typically become more stringent. Scientist trainees acquire a "sense" of what their discipline expects at its current state of development. The sense is fine-tuned by immersion in the work of others and frequent correction when statements are poorly supported. (This can, incidentally, be a fairly painful and confusing process.)

Consider this culturally evolving, individually learned, and collectively tuned sense of evidential support that exists within a scientific community. This is a normative intuition of proper concern for an epistemologist, and it presents the usual cluster of difficulties with norms. An individual considers stating a certain claim but is somehow bothered by it in that it seems overstated. This is the same sense that tells her that claims made by others are overstated. The propositional content of the intuitions is difficult to make out. One might try "the claim is overstated," but overstated with respect to what? With respect to the evolving set of standards of the discipline. So far, so good. But why, then, does the intuition seem to motivate directly rather than via some consideration of the extent to which one is committed to those standards or what one has to lose from their violation? The standard solution is to claim that the individual has a preference not to overstate her claim and that her belief that it is overstated makes not making the claim the rational thing to do. So, apparently, it is really rational to be moral. This just throws the normativity back on the rational, which needs an account in its own right and, in any event, tells us nothing about whether such preferences are good things to have. Perhaps, on the other hand, the content of the intuition is not in fact propositional (although there is again a great deal we can say about it propositionally) but primitive in just the way warning cries are primitive. The sense of overstatement has a culturally selective history which determines extension and intension for the sense itself, which accommodates its learned and evolving nature.

In sum, cultural flexibility constitutes no threat to the significance of primitive content or of the ability of naturalism to fully account for norms. Both responses, that our biological inheritance allows our flexibility and that the shorter-term functional histories also determine primitive content, press this point. The difference between the two schemas, however, concerns where the benefits of selection accrue. Where there are biologically selected function-stabilizing mechanisms behind our normative intuitions, we can expect them to be reasonably reliable with respect to factors bearing on long-term reproductive success, although they may be too general to cope well with current novelties. Cultural selection, while responding more quickly to environmental changes, may result in selective histories, function specification, and primitive content that serves only the proliferation of the cultural form itself, possibly at the expense of baseline human well-being. Consequently, for the purposes of epistemology, one should expect that the reliability in environmental tracking that we expect of science is maintained generally by inherited factors (e.g., detecting the failure of predictions) and within those constraints by the culturally evolved and individually learned standards of changing disciplines. Primitive content is the key to understanding the meaning and, thus, objectivity of standards of both sorts.

A BRIEF DIGRESSION: SUPERVENIENCE?

One of the ways in which philosophers have attempted to accommodate the meaning of normative intuitions within the propositional paradigm is to claim that normative intuitions do express propositions, but the properties that are attributed by those propositions are not natural but nonnatural properties (Kim 1988). With the ascendancy of the natural sciences, it has become unfashionable to maintain that there is anything wholly unrelated to matter and energy, so the notion of "supervenience" has been employed to attempt to account for nonnatural properties within the natural world. Consider the proposition "food is good." A nonnatural property "goodness" is being attributed to food. Some might claim that goodness is only nonnatural in the sense that it supervenes on ordinary physical properties such as mass, momentum, and electrical charge. To say that goodness supervenes on natural properties is to say that it is always some natural property or other that accounts for the goodness – there is no extra spiritual stuff to account for. But there may be all kinds of natural properties that might create goodness, so that goodness is not a natural *category* but a supervenient evaluative category. Along with the supervenience claim often comes the conclusion that supervenient categories are irreducible to natural properties since what gathers up all the natural bases for goodness is not some natural law but some other essentially evaluative set of standards. This is supposed to allow dispensing with nonnatural properties as separate items of ontology while allowing the disciplinary autonomy (i.e., blocking the disciplinary reduction) that nonnatural properties allowed.

Horgan and Timmons (1992) argue that while it is certainly possible that such supervenience relationships exist, the realist about normative properties must do more than simply defend the possibility of such relations. One must *explain* what determines which supervenience relation holds between truth-makers for normative and physicalistic language, "truth-maker" just being another word for extension. The approach taken in this book does not rely on the notion of supervenience to relate moral and physicalistic truth-makers, principally because the reification of moral truth-makers into moral *properties* seems unhelpful. But it may help clarify the relation of the present proposal to current alternatives to rephrase it in terms of the (to some) more familiar supervenience relations.

The most important characteristic of the supervenience relation is that the higher-level property supervenes on a *disjunction* of physical base properties. That is, the collection of different physical states that are sufficient for the instantiation of the supervenient or higher-level property may form a rather motley collection of various sorts of physical states with ad hoc restrictions and the collection may be governed by a nonsystematic array of physical laws. So, for instance, being a table is not a natural property but rather supervenes on various physical configurations without the addition of any mysterious element of table-ness. Still, it seems that to claim there is even a supervenient property of being a table, one needs to say something about how such a property gets defined, at least enough to give us reasonable grounds for thinking that there might be a *coherent* supervenient property. Obviously, this is going to be complicated, involving reference to use by human beings, along with certain historical facts about the application of the term "table." Presumably, the specifications will be disjunctive: this configuration or that configuration or that other configuration, and so on.

The challenge presented by Timmons and Horgan is to offer a systematic way of saying just what it is that collects the disjuncts together in the set that forms the subvenient base of the higher level property. We have already done this for functional semantics in the last chapter. Recall that the portion of the general rule for an adapted mechanism R_M whose failure (the various [**condition** & ¬**process**] configurations) forms the correspondence conditions for some corrective signal **cs** is, in fact, just such a disjunctive set as supervenience relations are invoked to accommodate. The corrective signal has as many distinct truth conditions as it has been selected for co-occurring with. More generally, the truth conditions for signals in adapted signaling can be expected to be disjunctive in physical terms, since the set of correspondence conditions (extension) for a given signal {**w** | **P** was selected for sending **s** in **w**} is determined by the historical efficacy of the adapted response to the signal, *not* by whether members of the set form a proper natural kind. Consequently, the biosemantic approach may be unique in its ability to create objective disjunctive truth conditions for normative utterances. If one can't resist the temptation to reify them into special sorts of disjunctive supervening properties, then the adaptive history of signaling systems provides a systematic factual basis for specifying the supervenience relations of the sort Horgan and Timmons insist on.

PROSPECTS FOR A THEORY OF OUGHT IN GENERAL

It is important to be as clear as possible about the limits of the primitive content proposal. We have seen that Hume and Moore's principled arguments do not succeed against it in the way they are supposed to, but there are nonetheless things that such a theory cannot do. First, as discussed earlier, naturalistic accounts of normative systems cannot participate in those systems in the way that normative theories have been designed to do. This limitation, I have argued, is not one of the incompleteness of the account, but simply of it being one kind of thing rather than another. The only kind of an account that can do more is one that is actually *part* of the system in question and thus can authoritatively participate in its activity.

Even if the functional characterization of primitive content does provide the correct semantic analysis of normative intuitions, some may claim that we have come no closer to answering the normative philosophical question, "what *ought* I to do?" The objection I am thinking of allows that normative intuitions are, in fact, more like animal warning cries than declarative sentences, and their truth has the kind of implications that warning cries have. If this is the case, the metaphysical and epistemological worries may have been answered. Nonetheless, this does not address the normative question itself because functional semantics only assesses the truth of particular normative intuitions. The normative question, on the other hand, requires understanding, not just the truth conditions of particular intuitions but pronouncing on the general issue of what one does when intuitions conflict, as they so often do.

As we learn more about the evolution and architecture of the human mind, one thing is becoming abundantly clear. The indivisible mind, the unified seat of consciousness, is a myth. No matter whether one prefers modern neuroscience set against an evolutionary backdrop or the psychology of conscious and unconscious mind, the human mind is composed of parts with different functions and is characterized by frequent conflicts between those parts. Both external analyses and our own internal experiences confirm this.

As a consequence, any theory that grounds normativity in the system stabilizing functions of particular mechanisms will be characterized by conflicts between particular sorts of normative standards implicit in the functional design of those mechanisms. For instance, if you are a vervet, a danger cry may truthfully and directly command flight and, at the same time, the voice of young offspring playing nearby may equally truthfully and directly demand that you stay and protect them. Presumably, one must make a decision. How does one resolve the conflict between equally true demands on one's behavior? The piecemeal analysis of the function of the danger-fleeing and offspring-protecting systems tell us nothing about what kinds of authoritative conventions might exist for adjudicating conflicts between regulatory subsystems. As a result, one might suspect that the functional analysis–primitive content approach to norms will turn out to be just another failed naturalistic attempt to deal with the philosophical question. But it is far too soon to draw such a conclusion.

In the first place, the primitive-content hypothesis predicts conflicts between competing normative systems and, despite the fact that philosophers have typically been concerned to find theories which resolve such conflicts, their existence does not constitute evidence against the hypothesis. On the contrary, the philosophical concern itself is due to the perennial existence of such conflicts. The most familiar, of course, is the conflict of the moral with the rational. Rationality requires that we act with maximal efficiency in satisfying our desires, given our beliefs about the world. Morality typically constrains this self-serving behavior, requiring us to make sacrifices that violate this constraint. This is not to say that moral imperatives and rational decisions always conflict but that it is at least plausible that part of the function of morality is to reduce the very efficiency in the satisfaction of desires that rationality demands. Theoretical approaches to normativity which presume a single kind of normativity, such as the philosophical question hopes for, have a difficult time dealing with the possibility of various incommensurable species of legitimate normative authority. The most they can hope to do is merely to *allow* that such species exist – that there are moral "oughts" and rational "oughts" and one can't get one from the other. The primitive-content hypothesis, on the other hand, actively predicts such incommensurable species and the seemingly irresolvable conflicts that arise from them. Thus, at least as an empirical hypothesis, primitive content gets it right.

The prediction of conflicts that seem irresolvable because of the mutual untranslatability of the underlying semantics does not, however, entail that such conflicts are as unresolvable as they seem. It does imply that, in particular cases, one may be unable to resolve a conflict from within either one or both of the conflicting subsystems. This does not mean that no resolution is to be found, only that if it exists, it must come from without.

Consider how we actually deal with normative conflicts, or, more to the point, how we judge the compromises of others. Inductive inference, for instance, violates the rules of valid deductive reasoning. Nonetheless, the standards of inductive inference are fully legitimate in their own domain. Rather than possessing a global normative system for reasoning, we seem to have several such systems combined with a sense of where each holds sway. Sometimes we do math and logic, sometimes we are learning about the messy world. Similarly, conflicts between the moral and the rational are adjudicated by our sense of which system holds sway in the circumstances, or due to the relative severity of violations. Suppose that in my haste to close a multimillion dollar deal, I swiped a disposable ballpoint pen from the desk of an absent stranger. Technically, the pen did not belong to me, and I had no right to take it. What I did was wrong. Still, we are likely to be forgiving in such a case since the rational motive was so strong and the moral violation so small. In other cases, we seem to feel the moral holds sway.

Typically, philosophers and decision theorists like to create systems that are not characterized by normative conflicts. Conflicts between the rational and the moral can be resolved via the "revealed preferences" approach to rational action, whereby the true preferences of an individual are revealed in action. So, if the individual sacrifices personal well-being to conform to some moral rule, we simply observe that the individual obviously had a strong preference for conforming to the rule. Thus, no compromise between reason and morality is necessary. Of course, this only solves the decision theorist's problem; the ethicist is left with the question of what makes such a preference legitimate. In the meantime, the decision theorist seems to have taken rationality from being a particular control system distinctive of human beings to an analytical framework that applies equally well to obviously nonrational plants and animals.

The resolution of conflicts offered by the primitive-content hypothesis has rather a different character – one more in line with how we actually resolve the conflicts. I have suggested that what we actually have and, thus, what we should be trying to understand is not an overriding rule structure which subsumes all of the particulars involved in various kinds of normative behavior, but, in the first place, a multiplicity of rule structures deriving from the roles of individual regulatory mechanisms and, second, a "sense" of when and in what circumstances each of the associated warning voices should be heeded first. Thus, if the primitive-content approach can offer any sort of answer to the general philosophical question (even as a mere description of the semantics), it will concern the nature and function of our sense of which of the various sets of standards which bind us hold sway in which circumstances.

We need not pursue the question at any length here, for the analysis of such "higher-level" normative systems is, at least in the abstract, no different from the analysis of any other. One simply asks what the function of this governing sense of applicability is, how it acquired that function, and how it operated while it was acquiring the function. The functional history supplies us with primitive-content semantics for the sense, which in turn allows us to give the descriptive answer to the philosophical question. What ought you to do, in general? What is true is that you ought to obey the particular (true) oughts that apply to your particular situation, where the very existence of such a meaningful higher-level set of applicability norms implies that most people pretty much *know* which rules apply, most of the time. Once again, this specification of the truth of the answer to the question falls short of actually telling you what to do, but it seems to me that if you are asking the question in good faith, you will be able to figure out what to do.

We should remind ourselves at this juncture that it makes little difference whether our sense of applicability of the various species of normative standards is instinctive or learned – the product of genetic selection, sociocultural selection, or a mixture of both. Either functional-selective history can provide primitive content for this sense. The interesting questions have all become empirical ones. How instinctive are the various specific systems of rational, epistemic, and moral norms? How instinctive is the overriding sense of applicability of the particular normative systems? And how *much* more complex is the actual hierarchical regulatory-normative structure than has been indicated here? These are all empirical questions. Moreover, despite the usefulness of treating morality and rationality as simple and unified normative subsystems for the purpose of discussing conflicts, in reality we are likely to find tremendous variation in the degree of instinctiveness of all sorts of normative intuitions, copious internal conflicts within the various subsystems, and considerable sharing of cognitive resources between the various systems. Nonetheless, it seems to me that understanding primitive content in hierarchical regulatory systems dispels the very old philosophical mysteries and can provide a needed focal point for a new empirical investigation into the nature of normative legitimacy as well.

Epilogue

Paley's Watch and Other Stories

Walking along a rocky beach, you notice that rocks and pebbles of different sizes have been arranged according to size in neat bands by the mindless interaction of waves and shore, the ordering effect of differential stability of material arrangements that is no more, and no less, than natural selection itself. A little farther down, you crouch by a tide pool and find, of all things, an old-fashioned pocket watch lying among the seaweed and anemones. The seaweed and anemones are easy enough to understand – natural selection along with the effects of heritable variation, and a *lot* of time, suffice to account for the lifeforms. The watch is a bit more of a puzzle. Perhaps William Paley has been by, seeding the beach with watches, trying to get us to think, proving the existence of God. That would certainly account for it.

Of course, it's not enough to merely propose a hypothesis that *would* account for the data, if it were true. Paley has been dead for almost two hundred years, and although it is possible that his spirit roams the earth dispensing watches, that's probably *not* where the watch came from, however well that would explain its presence.

Paley asked us to consider whether, finding a watch on the beach, we would note its organized complexity and fitness to its task. From this observation, would we infer that it was the result of random physical processes, or would we infer that it was the product of intelligent design? If we make the inference to intelligent design in the case of the watch, should we not be even more compelled to do so when we find biological organisms, which exceed the watch in both organized complexity and fitness to task?

Paley's argument, along with all the other versions of the "argument from design," is often characterized as an argument by analogy. The reasoning is supposed to be that if organisms are like watches in their organized complexity, then like watches they imply intelligent design. It does not take much analysis to see, however, that the argument from design rests on a rather poor analogy, watches and organisms differing in far more particulars than they resemble one another. But some philosophers have argued that Paley's argument was in fact not analogical, but instead an inference to the best explanation. On this construal, the watch serves merely to evoke the pattern of reasoning, rather than the particular features of biological organisms. The hypothesis of God predicts the existence of organized complexity much more strongly than the hypothesis of random physical forces; therefore, the existence of such organisms favors existence of God. Moreover, as an inference to the best explanation, Paley's argument was perfectly good – in 1805. There simply was no hypothesis, other than the existence of an intelligent creator, that *could* plausibly explain the observed order of nature at that time.

If Paley's watch is one of the metaphorical touchstones of empiricist philosophy, Otto Neurath's boat is another. Neurath likened conceptual progress to rebuilding a ship on the ocean while traveling in it. We replace only one plank at a time, with care, and only when the risks involved in the repairs are outweighed by the risks of not making them. The image of a ship at sea emphasizes two important aspects of conceptual innovation. First, we cannot get out of our worldview and haul it out in dry dock while changes are made. Changes must be made while we are in it, despite the discomfort and inconvenience this may cause. Second, it is easy to understand why people get worried when repairs are suggested. Holes in boats have a nasty way of letting in water, which is the last thing one wants to happen. People resist conceptual innovation because they can, because – unlike the ugly truth – no one can force you to accept a revised worldview, and because they are quite rightly concerned with the possibility that the new revolution might actually make things worse.

The subject matter of this book – human beings as evolved material beings, essential norms of reason and social behavior as conventions of varying antiquity and authority – are just the sort of innovations that make many people nervous. Many people still do not accept the accumulating evidence for our very long evolutionary history because it makes them nervous, and because they don't have to. Even for those who do, mixing the scientific and the normative violates taboos entrenched in long history.

Paley's inference to the best explanation was a valid inference of that sort due to the availability, or unavailability, of coherent alternative theories of the origin of order in the world. All of that changed with Darwin – not because Darwin proved anything one way or another regarding the existence of a deity, but because presentation of a viable alternative can change what

Epilogue

the best explanation is. But neither Darwin nor the legions of scientists who have been tirelessly filling in the details, both theoretical and factual, of the evolutionary story can convert those who wish not to be converted. Inductive reasoning is like that.

Arguments concerning the limitations of naturalism with respect to the normative resemble Paley's in that they depend critically on certain assumptions concerning the field of possible theories. Change the field in any significant way and those arguments need to be reevaluated, at the very least. The rather modest ambitions for Part III of this book were simply to make the case that the field has changed. Functional semantic theories have been around for a while and, while Millikan's highly developed version has yet to win general approval (or even broad notice outside of philosophy), its presence and coherence are a fact that cannot be denied. My own contribution is simply to point out that in its simplest version, it converges with work in the evolution of meaning conventions and allows conventional (as opposed to natural) meaning for signals that both track and motivate, which have truth conditions and normative behavioral consequences. These are the features of normative deliberation that have proven impossible to give full accounts of in familiar terms, and it is only that difficulty that legitimates the insistence that the normative is of a world apart. The mere presence of a coherent alternative changes the game entirely.

Part III was directed not toward those who will fight the encroachment of scientific inquiry into philosophical territory on every front, but toward those who might welcome such an account but have heard or been taught that no naturalistic account of the normative can be correct. I have no illusions concerning the difficulty of establishing naturalistic theses of this kind conclusively and make no pretense to have done so. Nor do I have any illusions about my own ability to canvass all objections that have ever or might ever be made to such a theory. But what one can do is clarify the theoretical options, and I think that I have shown something that few have suspected is possible: that natural relationships exist which largely have the form we believe the objective rules of reason and behavior have. The game has changed not because of anything I have done, but because evolutionary theory exists, and functional semantics exists, and it is just a matter of time before a well-developed and defended account of primitive content (or whatever it ends up being called) takes its place as a major, and perhaps sole, contender for the naturalist account of normative truths. Similarly, it is beyond my power to make everyone listen to and comprehend the theory but, by all rights, one cannot continue to assess the prospects of naturalism according to the failings of early-twentieth-century emotivism. Better theories are possible; everything else is a straw man.

In a similar vein, I am aware that the models of Part II are only first steps toward what a rigorous evolutionary epistemology should look like, but I hope I have shown that the economic, selectionist, game-theoretic approach is the one that has to be taken, if we are to extend existing disciplines like cognitive science and evolutionary psychology to address the epistemological questions of how our concepts relate to the world, and how our beliefs track it. We cannot simply compare our concepts to the world, for we can only look at the world in terms of them. So we need theoretical tools that help us understand what sorts of concepts stabilize in a heterogeneous but unanalyzed world. Natural selection is simply the summation of the effects of stabilizing agencies. Game theory is just strategic economic analysis. The reason quantitative analysis is necessary is because that is how scientific theory is done – not with metaphor. The reason selection models are necessary is because selection is what forges the bonds between mind and the world. That, I think, is *how* we need to proceed. What we may discover depends on how the adventure unfolds.

Ursula Le Guinn's delightful *Earthsea Trilogy* involves a concept of magic which requires that one know the *true name* of a thing to gain mastery over it. Each thing was given a name in the making of the world, and the sorcerer who learns these names can command the named to do his bidding. In the real world, philosophers of science often claim that an ideal set of scientific concepts names "natural kinds" or "cuts nature at its joints." But just as there is no reason to think that our world consists of things which were given names in its making, the knowledge of which confers mastery, there is no reason to think that the world comes in packages just the right size to fit our conceptual capacities, at least without appeal to a beneficent and omnipotent creator. Unfortunately for epistemology, evolution does not optimize, it *satisfies*. Nature selects the best of the alternatives life presents to it and guarantees only that the survivors are good enough to survive. That is what we are, good enough; better than most, perhaps, but mostly just not-dead-yet. Thus, the quality of our knowledge can only be understood against the backdrop of our continued existence. If our choices never mattered, then we would know nothing and our thoughts would be without meaning. But our choices do matter, and the continuation of each human limb of the tree of life depends absolutely on the wisdom of those choices.

Epistemology matters because knowledge is not a simple thing. Because evolution satisfices, because things do not have true names, and because reason is but one of the ideals we pursue, knowledge is also a relative thing.

Epilogue

But the mere fact of relativity does not mean that anything goes. For a given task, some tools are better than others. General tools, like our system of scientific knowledge, can take different forms because of the large number of conventions involved in their formation and because of their contingent histories. But their construction and utility is systematic, and this construction *can* be understood, even if we cannot ever get outside of it.

Notes

CHAPTER 1

1. I should emphasize that the notion of a "replicator" and its shortcomings which I discuss in this chapter only peripherally concern the "replicator dynamics" (Schuster and Sigmund 1983). Although the family of formalisms that Schuster and Sigmund generalize as the "replicator dynamics" does not presuppose the existence of the more narrowly defined replicators discussed in this chapter, these replicators do in many cases behave according to a related dynamic (thus the name). In contrast, what the replicator dynamics does presuppose is that certain indexes, which may (but need not) be construed as measures of the relative frequency of tokens of certain types, change in conformity with the specified dynamic. This defines a more general class of processes, with which we are concerned in Part II.
2. I use the term *Darwinian process* to refer, loosely, to processes exhibiting functional improvements due to undirected variation and systematic environmental selection.
3. Note that redundancies in the genetic code allow functional equivalence without precisely identical chemical structures. It is usually practical to individuate genes in terms of their decoding consequences, rather than their precise chemical structure.
4. See Aunger (2000) for a variety of views on this point.
5. This statement is probably the clearest expression of the assumption that Darwinian processes need replicators, since he infers that the lack of replicators would keep "Darwinism" from happening.
6. There is considerable potential for confusion concerning the use of "levels of selection" in applications of evolutionary theory. In particular, the "levels" in the group selection debate are levels of scale, whereas the levels with which we are concerned are roughly the levels of a regulatory hierarchy.
7. This will be no surprise to those who are familiar with Hull's work on the metaphysics of biological lineages. See especially Ghiselin (1997).
8. Incidentally, Rosenberg (1992) brings up similar problems with Hull's definitions, particularly with the type-token aspects.
9. Hull (1988), 241, 244, 377, 404, 406, 407, 412, 424, 458, 513.
10. In all fairness, this is not Hull's final word on the subject. See Hull (2001).
11. The notion of selfishness in evolution is discussed in the next chapter.

12. Information, and in particular the probability measures used in information theory, are discussed at length in Chapter 4.
13. Try Dahlbom's (1993) *Dennett and His Critics* for starters.
14. I especially like Sperber (1996) on this point.
15. I have my say about how "self-replication" should be understood in the next chapter.

CHAPTER 2

1. At this point, some readers may be puzzled as to the philosophical orientation behind this project for, on one hand, my comments about metaphysics indicate a skepticism regarding our ability to get at the underlying nature of reality in some final way and a pragmatism about the object of theory building. On the other hand, my dissatisfaction with intentional or relational characterizations of higher level (i.e., cultural or multimolecular) entities indicates a commitment to a sort of materialist reductionism that is most commonly associated with realism or even foundationalism. The connection between the two is just the conviction that if one seriously wants to contribute philosophically to the *advancement* of the body of human knowledge, as opposed, say, to defending some existing practice or theoretical framework, then reduction is what one attempts to do, and theoretical entities which make reduction more difficult may need to be dispensed with.
2. Cf. Sober and Wilson (1998) on individualistic biases in evolutionary theory and psychology.
3. I am indebted to Hahlweg and Hooker's (1989) synthesis of Waddington's developmental biology and Piaget's developmental psychology for the account given here.
4. See Bonner, *The Evolution of Culture in Animals* (1980).
5. One concern that sometimes comes up is whether imitation requires complex intentionality, thus excluding phenomenon such as birdsong transmission from the proper domain of memetics. My view on this is that while deliberate imitation is certainly a more complex process than reflexive imitation, I don't see what difference this makes for the application of the replicator concept. If anything, unconscious imitation may provide *better* examples of selfish replicators.

CHAPTER 3

1. Specifically, the Hardy-Weinberg and Fisher equations. See Hofbauer and Sigmund (1988).
2. Notice that the loss to mutation is multiplied by and therefore proportional to the frequency. I discuss this presently.
3. Belief individuation constitutes a challenging theoretical topic which I do not intend to contribute to here. What matters for our purposes is that we can expect to be able to count beliefs in a population if we need to. Sperber (1996) has argued persuasively that this is easier than one might suspect.

Notes to pp. 109–137

CHAPTER 4

1. Reprinted in Shannon and Weaver (1949).
2. So, for instance, $\log_2 2 = 1$; $\log_2 4 = 2$; $\log_2 8 = 3$; $\log_2 16 = 4$; etc.
3. This is the aspect of information theory on which Dretske (1981) focused, discussed later.
4. Dretske's (1981) extraction of the single term from the rate of transmission formulation differs. He had it that the information in $\mathbf{R}_j$ about $\mathbf{S}_i$ was

$$I(\mathbf{R}_j; \mathbf{S}_i) = -\log_2 \Pr(\mathbf{S}_i) + \sum_j \Pr(\mathbf{S}_i \mid \mathbf{R}_j) \log_2 \Pr(\mathbf{S}_i \mid \mathbf{R}_j),$$

   or the information generated by $\mathbf{S}_i$'s occurrence minus the uncertainty about $\mathbf{S}_i$ averaged over all states of $\mathbf{R}$. The latter part is a bit of a puzzle. Why average over all states of $\mathbf{R}$?
5. The alternative formulations will become relevant to Godfrey-Smith's discussion of reliability measures, discussed later.
6. The notion of freedom of choice used here is something to worry about. On this construal, you are more free if the likelihood of your choosing one of two options is 50 percent than if it is, say, 75 percent. The counterexamples are obvious. I offer you $100 without obligation. If you act freely, it is virtually certain that you will accept.
7. There is nothing surprising about this additivity. Logarithms are designed to turn multiplication (e.g., of independent probabilities) into addition (e.g., of the associated information).
8. See, however, Kapur (1994) for an "unorthodox" information measure that shares some of the desirable properties of Shannon's entropy.
9. To compute base 2 logs on your calculator: $\log_2 x = \ln x / \ln 2$.
10. Godfrey-Smith (1996) aptly dubs the distinction "cartesian" versus "Jamesian" reliability.
11. This was verified via both graphical and computational analysis. An algebraic proof for $\mathbf{n} = 2$ may be possible.
12. Rates calculated over runs of one million trials each.
13. Harms (1996a) (appendix) includes an analytic proof of information gain under selection, on an alternative formulation. See also Harms (1997) for an application of mutual information in an evolutionary learning model.
14. My focus here has been on applications of information theory to naturalistic epistemology. I would be remiss, however, were I not to mention some interesting work by Elliott Sober and Martin Barrett applying mutual information to the analysis of temporal asymmetries in Bayesian inference. See Sober (1991), Barrett and Sober (1992), Sober and Barrett (1992).

CHAPTER 5

1. Please note that I am using the term *reality* to mean Kant's noumenon, not the totality of the phenomenal world, as the term is sometimes used.
2. Godfrey-Smith (1996) has also been pursuing this line of thought.

Notes to pp. 152–169

CHAPTER 6

1. Many readers doubtless will be disturbed by my use of terms such as *well-being*, *human flourishing*, and what is *good for* people in the context of evolutionary discussions. To such worries, I would simply say that if you step back a bit and think about what successful multigenerational reproduction involves for human beings, including especially the environmental requirements for protecting, raising and teaching our children, most if not all of the factors involved in well-being and flourishing can be accounted for as precisely the requirements for "raising the next generation to raise the next generation." Human reproductive fitness is not just a matter of producing more offspring than anyone else, especially outside of the protections and subsidies of modern civilization. Individual health, prosperity, and the respect of others, as well as a safe, stable, and culturally rich child-rearing environment, are all important factors in raising offspring with the best chance of successfully raising their own.
2. These are also a subset of what sociobiologists call "epigenetic rules" – genetically determined guidelines that govern either development or behavior. See Alexander (1990).
3. Notice that here as always, the maximum information as well as the current information is relative to the manner of partitioning the state space of the environment.
4. Notice again that this maximum information level is determined by the partition on the space of the environment.
5. Campbell (1974) describes similar systems at the lower levels of his "hierarchy of blind variation and selective retention processes."
6. Notice that if the high frequency of a certain belief were a sure indication that not having it was more fit_G, then we get information about the world held in a fashion in which it is counterproductive. Thankfully, such pathological cases should disappear quickly under the force of selection on organisms.
7. See Philip Kitcher's *Vaulting Ambition* (1985) for an extensive discussion of these issues.

CHAPTER 7

1. Montague et al. (1995) implicitly make this assumption as well.
2. Notice again that in setting up a model like this, it is not always obvious what should count as the "same" token. In this case, when a bee starts the foraging cycle over again, we must choose whether to call it the same behavior token with a new type or a new behavior token generated to fill the "niche" of the old. In this case characterizing the tokens as the same, and the process as that of mutation, is convenient. In some other cases, ontology may constrain such choices on the part of the modeler.
3. It may seem odd to apply evolutionary theory to constant-size collections of tokens, but this is a standard feature of the kind of models now referred to as "genetic algorithm" models. See Holland (1992).
4. For simplicity, I have imagined that full flowers take twice as long as empty ones, although in most real cases it will take considerably more time than that. The important feature for the dynamics is that it takes a significant amount of time

to go to an empty flower, and this constrains the efficiency of the random-search strategy.
5. Börgers and Sarin (1994) have made some headway in linking individual stochastic learning-through-reinforcement dynamics with population-level replicator dynamics. Recall also that Montague et al. have proposed a detailed neurological model of this sort of process, which is compatible with the treatment here.
6. Notice that I am not assuming that the individual bees go to flowers according to the relative strength of their preferences for blue and yellow, that is, I am not assuming that they are "probability matching." See Siegel (1961) for the basic literature.
7. See Hofbauer and Sigmund (1988), Chapter 1.
8. Incidentally, to simulate partial accuracy in the selection on preferences π, one merely needs to lower the selection intensity **I**. The interpretation on the individual level may be different, but the result on the population level (i.e., the expectation of the errors) is the same.
9. Note that I am not here claiming that selection is *for* information but that it is at least selection *of* higher information.

CHAPTER 8

1. Deacon (1997) persuasively makes the case for the peculiarity of the human representational system, as well as pointing out the problem with taking it as basic.
2. Note that evolutionizing this game without signal costs assumes that the cooperative problem has been solved, an assumption that has become increasingly reasonable in the last twenty years or so (Sober and Wilson 1998).
3. Millikan (1984) used the term *intentional icon* for this kind of signal, reserving the term *representation* for signals with interchangeable parts, such as pictures and sentences. She has since started using "representation" in this wider sense.
4. See Millikan (1990) for a comparative discussion of ways of specifying reference.
5. Cf. Millikan, "Pushmi-Pullyu Representations" (1996), for a discussion of this point.
6. While the consensus seems to be that animal communication lacks combinatorial syntax and thus any analog to words, their ability to identify individuals and learn associatively strongly suggests that their internal "thought" processes have somewhat more structure.
7. See especially Axelrod (1984).

CHAPTER 9

1. "Natural" meaning is usually understood to be the kind of statistical indication that causal traces have, like the way the bear's footprint "means" that a bear was here.
2. Moore (1903); Darwall, Gibbard, and Railton (1992), 115–21.
3. Thomas Nagel (1974) may be blamed for starting this "what is it like" business in his famous article "What Is It Like to Be a Bat?"

Appendix

Proof of Information Gain under Frequency-Independent Discrete Replicator Dynamics for Population of n Types

I consider information gain in fixed environments, in discrete time, for populations of n types. "Fixed environments" means fixed fitnesses, so that unlike most evolutionary models, the fitnesses of types are not dependent on variations in frequencies of the other types in the population which form part of their environment. The justification for this is that, if we are looking for information about the environment in population distributions, we need to begin with a case in which the external environment is the only force that affects the evolution of the population, excluding the complex and potentially chaotic effects of the internal dynamics of frequency-dependent fitnesses.

Let $\vec{x} = \langle x_1, \ldots, x_n \rangle$ be the vector of relative frequencies of types $i \in \{1 \ldots n\}$, such that for all types i, $x_i \geq 0$, and $\sum x_i = 1$, and let $\vec{w} = \langle w_1, \ldots, w_n \rangle$ be the vector of type fitnesses ("the environment") such that for all types $i \in \{1 \ldots n\}$, $w_i \geq 0$. The standard discrete time replicator dynamics specifies the new frequencies of each type according to the old frequencies and the fitnesses, according to

$$x_i^I = \frac{x_i w_i}{\sum_j x_j w_j}.$$

For our purposes, it will be more useful to specify type frequencies as a function of time and some initial distribution. The equivalent time-dependent formulation $x_i(t) = x_i w_i^t / \sum_j x_j w_j^t$ is determined by induction on t.

(1) If x_i is the frequency of some type i at $t = 0$, then $x_i(0) = x_i / \sum_j x_j = x_i w_i^0 / \sum_j x_j w_j^0$, and (2) the frequency of type i at $t = 1$ will be $x_i(1) = x_i w_i / \sum_j x_j w_j = x_i w_i^1 / \sum_j x_j w_j^1$. (3) If at some time t the type frequencies

are $x_i(t) = x_i w_i^t / \sum_j x_j w_j^t$, then at $t+1$, the frequencies will be

$$x_i(t+1) = \frac{x_i(t) w_i}{\sum_j x_j(t) w_j} = \frac{\frac{x_i w_i^t}{\sum_k x_k w_k^t} \cdot w_i}{\sum_j \frac{x_j w_j^t}{\sum_k x_k w_k^t} \cdot w_j} = \frac{\frac{x_i w_i^{t+1}}{\sum_k x_k w_k^t}}{\sum_j \frac{x_j w_j^{t+1}}{\sum_k x_k w_k^t}}$$

$$= \frac{x_i w_i^{t+1}}{\sum_j x_j w_j^{t+1}}.$$

Consequently, for any time $t \in \mathbb{N}$, $x_i(t) = x_i w_i^t / \sum_j x_j w_j^t$.

Mutual information is a measure of statistical correlation between two systems, so we need to define partitions on the space of the population distribution and the environment, and we need to define a probability measure over the joint space.

Let $\mathbf{X}$ be the simplex state space of the distribution vectors $\vec{x}$, Δ^{n-1}, and let $\mathbf{W}$ be the state space of the fitness vectors $\vec{w}$, $\mathbb{R}^n_+$. Then $(\vec{x}, \vec{w})$ are points in the joint space $\mathbf{X} \times \mathbf{W}$. Partitions will be defined on this joint space: $\mathbf{S}_i \subseteq \mathbf{X} \times \mathbf{W}$, $\mathbf{S} = \{(\vec{x}, \vec{w}) | \forall j \neq i, w_i > w_j\}$ are states of the environment characterized as the state where type i is the most fit. Similarly, $\mathbf{D}_i \subseteq \mathbf{X} \times \mathbf{W}$, $\mathbf{D}_i = \{(\vec{x}, \vec{w}) | \forall j \neq i, x_i > x_i\}$ are states of the population distribution where type **I** is the most frequent. States $\mathbf{S}_0$ and $\mathbf{D}_0$ where two or more types are equally most fit or frequent (respectively) will be disregarded in the following, since they will turn out to have probability zero on the measure we will define and so will not affect the information results.

Pr_0 is some probability measure defined on the joint space $\mathbf{X} \times \mathbf{W}$ which has the following properties. (1) Pr_0 is a uniform Lebesgue measure over each subspace $\mathbf{X} \times \vec{w}$, (2) Pr_0 is invariant over permutations in types. (The nature of this invariance is made exact in what follows.) The latter permutation invariance requires a sort of symmetry over the joint space and is a weaker requirement than some broader uniformity requirements one might impose.

The symmetry of Pr_0 over the joint space will be exploited for the purposes of the proof via a two-place permutation function: $f_{j,k} : \mathbf{X} \times \mathbf{W} \Rightarrow \mathbf{X} \times \mathbf{W}$ or $f_{j,k} : (\vec{x}, \vec{w}) = (\vec{x}', \vec{x}')$ such that $\forall i \neq j, k, x_j^t = x_k, x_k^t = x_j, x_i^t = x_i$, $w_j^t = w_k, w_k^t = w_j$, and $w_i^t = w_i$. Intuitively, for any point $(\vec{x}, \vec{w})$, this simply swaps the jth and kth places in both $\vec{x}$ and $\vec{w}$, leaving the other places untouched.

$f_{j,k}$ applies to subsets s as follows: for $\sigma \subseteq \mathbf{X} \times \mathbf{W}, f_{j,k}(\sigma) = (\sigma') = \{f_{j,k}(\vec{x}, \vec{w}) | \vec{x}, \vec{w} \in \sigma\}$. The invariance of Pr_0 over permuations of types is just the requirement that $\text{Pr}_0(f_{j,k}(\sigma)) = \text{Pr}_0(\sigma)$.

Appendix

$\Pr_t$ is defined in terms of $\Pr_0$ via the version of the replicator dynamics defined as a function of time as given earlier. First, $\Pr_t(S_i) = \Pr_0(S_i)$ for all t and i, since the fitness does not change. Second, notice that

$$x_i w_i^t > x_j w_j^t \text{ iff } x_i w_i^t / \sum_k x_k w_k^t > x_j w_j^t / \sum_k x_k w_k^t$$

(i.e., the denominators can be dropped from the inequality since they are equal), so that

$$\Pr_t(\mathbf{D}_i) = \Pr_0(\forall j \neq i, x_i(t) > x_j(t)) = \Pr_0\left(\forall j \neq i, x_i w_i^t > x_j w_j^t\right).$$

The replicator dynamics are expressed in $\Pr_t$ via this definition.

Theorem: *For $\Pr_t$, S_j, and D_i as defined earlier, the average mutual information*

$$I_t(S, D) = \sum_{i=1}^n \sum_{j=1}^n \Pr_t(S_j \,\&\, D_i) \log_2 \frac{\Pr_t(S_j \mid D_i)}{\Pr_t(S_j)}$$

increases monotonically with $t \in \mathbb{N}$.

Proof: [(1)–(4) are of the nature of lemmas in support of the main argument in (5).]
(1) For any time $t \in \mathbb{N}$ and for all $i \in \{1 \ldots n\}$, $\Pr_t(\mathbf{D}_i \,\&\, \mathbf{S}_i) < \Pr_{t+1}(\mathbf{D}_i \,\&\, \mathbf{S}_i)$:
(1.1) If $((\vec{x}, \vec{w}) \in (\mathbf{D}_i \,\&\, \mathbf{S}_i)_t = \{(\vec{x}, \vec{w}) \mid (\forall l \neq i, x_i w_i^t > x_l w_l^t) \,\&\, (\forall l \neq i, w_i > w_l)\}$, then it will also be a member of the corresponding set at $t+1$, since if $w_i > w_l$ and $x_i w_i^t > x_l w_l^t$, then $x_i w_i^{t+1} > x_l w_l^{t+1}$. This shows that

$$\Pr_t(\mathbf{D}_i \,\&\, \mathbf{S}_i) \leq \Pr_{t+1}(\mathbf{D}_i \,\&\, \mathbf{S}_i).$$

(1.2) To show that the new probability is strictly greater, we need to show that there is a region of positive measure such that for every point $(\vec{x}, \vec{w})$ in the region,

(a) $\forall l \neq i \; w_i > w_l$, (the point $\vec{x}, \vec{w}$ is indeed an element of $\mathbf{S}_t$ all times t),
(b) $\exists l \neq i \; x_i w_i^t < x_l w_l^t$, $((\vec{x}, \vec{w})$ is not an element of $\mathbf{D}_i$ at t), and
(c) $\forall l \neq i \; x_i w_i^{t+1} x_l w_l^{t+1}$, $((\vec{x}, \vec{w})$ is an element of $\mathbf{D}_i$ at $t+1$).

Such a region can be specified as follows: for any $\vec{w}$, let w_α designate the largest component and let w_β designate the second largest component. Then we can specify a subset $\sigma_{\vec{w}} \subseteq X \times \vec{w}$ s.t.

$$\sigma_{\vec{w}} = \{(\vec{x}, \vec{w}) \mid x_\alpha + x_\beta > 2/3 \,\&\, x_\beta(w_\beta/w_\alpha)^{t+1} < x_\alpha < x_\beta(w_\beta/w_\alpha)^t\},$$

where x_α and x_β are the initial frequencies of the first and second most fit types. The first inequality ensures that either type α or type β is the most frequent at all times, the second ensures that β is the most frequent at t, and α is the most frequent at $t+1$. Since Pr_0 is uniform over each subspace $\mathbf{X} \times \vec{\mathbf{w}}$, $\sigma_{\vec{w}}$ has positive measure for each choice of $\vec{\mathbf{w}}$. Consequently, the union of the relevant subsets $\sigma_i = \cup \{\sigma_{\vec{w}} \mid \forall_{l \neq t}\ w_i > w_l\}$ has positive measure on the whole space $\mathbf{X} \times \mathbf{W}$, and each $(\vec{\mathbf{x}}, \vec{\mathbf{w}}) \in \sigma_i)$ satisfies (a) – (c), as previously noted.

(2) For any time $t \in \mathbb{N}$ and for all $i \in \{1 \ldots \mathbf{n}\}$, $\text{Pr}_t(\mathbf{S}_i) = \text{Pr}_{t-1}(\mathbf{S}_i)$, by the assumption of fixed fitnesses, as previously noted.

(3) For any time $t \in \mathbb{N}$ and for all $i \in \{1 \ldots \mathbf{n}\}$, $\text{Pr}_t(\mathbf{D}_i) = \text{Pr}_{t+1}(\mathbf{D}_i)$: For all i and t, $\text{Pr}_t(\mathbf{D}_i) = \text{Pr}_0(\forall_{l \neq i}\ x_i(t) > x_l(t)) = \text{Pr}_0(\forall_{l \neq i},\ x_i w_i^t > x_l w_l^t)$. Let $\sigma = \{(\vec{\mathbf{x}}, \vec{\mathbf{w}}) \mid \forall_{l \neq i},\ x_i w_i^t > x_l w_l^t\} = \mathbf{D}_i$. Then $f_{i,j}(\sigma) = \sigma' = \{(\vec{\mathbf{x}}, \vec{\mathbf{w}}) \mid \forall_{l \neq j},\ x_j w_j^t > x_l w_l^t\} = \mathbf{D}_j$. But since $\text{Pr}_0(\sigma) = \text{Pr}_0(\sigma')$, then for all i, j, t, $\text{Pr}_t(\mathbf{D}_i) = \text{Pr}_t(\mathbf{D}_j)$, and since $\sum_i \text{Pr}_t(\mathbf{D}_i) = 1$, it follows that $\text{Pr}_t(\mathbf{D}_i) = \text{Pr}_{t+1}(\mathbf{D}_i) = 1/\mathbf{n}$. It also follows from (1) and (3) and the definition of conditional probability that

(3.1) $$\text{Pr}_t(\mathbf{S}_i \mid \mathbf{D}_i) < \text{Pr}_{t+1}(\mathbf{S}_i \mid \mathbf{D}_i).$$

(4) $\text{Pr}_t(\mathbf{D}_i\ \&\ \mathbf{S}_j) = \mathbf{Pr}_t(\mathbf{D}_i\ \&\ \mathbf{S}_k)$ for all t and for all $j,k \neq i$:
Since $\mathbf{Pr}_t(\mathbf{D}_i) = \sum_j \mathbf{Pr}_t(\mathbf{D}_i\ \&\ \mathbf{S}_j)$ remains constant for all t by (3), and by (1) $\mathbf{Pr}_t(\mathbf{D}_i\ \&\ \mathbf{S}_i)$ increases with t, it follows that $\sum_{j \neq i} \mathbf{Pr}_t(\mathbf{D}_i\ \&\ \mathbf{S}_j)$ must decrease. Moreover, the terms of the latter sum must decrease uniformly. For all t, i, j,

$$\text{Pr}_t(\mathbf{D}_i\ \&\ \mathbf{S}_j) = \text{Pr}_0((\forall_{l \neq i},\ x_i w_i^t > x_l w_l^t)\ \&\ (\forall_{l \neq j},\ w_j > w_l)).$$

If we let $\sigma = \{(\vec{\mathbf{x}}, \vec{\mathbf{w}}) \mid (\forall_{l \neq i},\ x_i w_i^t > x_l w_l^t)\ \&\ (\forall_{l \neq j},\ w_t > w_l)\}$, then $f_{j,k}(\sigma) = \{(\vec{\mathbf{x}}, \vec{\mathbf{w}}) \mid (\forall_{l \neq i},\ x_i w_i^t > x_l w_l^t)\ \&\ (\forall_{l \neq k},\ w_k > w_l)\} = (\mathbf{D}_i\ \&\ \mathbf{S}_k)_i$. But since $\text{Pr}_0(\sigma) = \text{Pr}_0(\sigma')$, then for all t and all $j, k \neq i$,

$$\text{Pr}_t(\mathbf{D}_i\ \&\ \mathbf{S}_j) = \text{Pr}_t(\mathbf{D}_i\ \&\ \mathbf{S}_k),$$

and it also follows from the definition of conditional probability, and (3) that for all $j, k \neq i$,

(4.1) $$\text{Pr}_t(\mathbf{S}_j \mid \mathbf{D}_i) = \text{pr}_t(\mathbf{S}_k \mid \mathbf{D}_i).$$

(5) The information result depends on the behavior of the conditional entropy

(5.1) $$\sum_j \text{Pr}_t(\mathbf{S}_j \mid \mathbf{D}_j) \log_2 \text{Pr}_t(\mathbf{S}_j \mid \mathbf{D}_t)$$

over time. For any i, if we let $q = \text{Pr}_t(\mathbf{S}_i \mid \mathbf{D}_i)$, then by (4.1) the conditional

Appendix

entropy is equivalent to

(5.2) $\quad q \log_2 q + \sum_{j \neq i} ((1-q)/(n-1)) \log_2((1-q)/(n-1))$

$\quad = q \log_2 q + (1-q) \log_2((1-q)/n-1))$

$\quad = [q \log_2 q + (1-q) \log_2(1-q)] + (1-q) \log_2(1/(n-1)).$

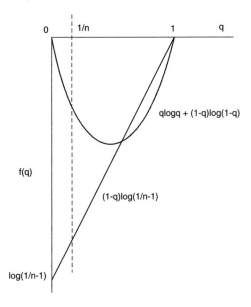

Now (5.2) is the sum of a linear function of q with a negative slope $(1-q) \log_2(1/(n-1))$ and a well-known parabolic function with continuously increasing slope, $q \log_2 q + (1-q) \log_2(1-q)$. The composite function must then have a continuously increasing slope, which entails that its value increases monotonically with q from its minimum value. The conditional entropy (5.1) reaches its minimum value when each of the conditional probabilities involved is equal, which is at $t = 0$ because of the charactersitics of Pr_0 as previously defined. From there, by (3.1), q increases monotonically with t, and consequently so does the conditional entropy. But since $\mathbf{Pr}_t(\mathbf{S}_i)$ and $Pr_t(\mathbf{D}_i)$ are fixed over t (by (2) and (3)), the "rate of transmission,"

$$I_t(\mathbf{S}, \mathbf{D}) = \sum_{j=1}^{n} -\Pr_t(\mathbf{S}_j) \log_2 \Pr_t(\mathbf{S}_j)$$

$$+ \sum_{i=1}^{n} \Pr_t(\mathbf{D}_i) \sum_{j=1}^{n} \Pr_t(\mathbf{S}_j \mid \mathbf{D}_i) \log_2 \Pr_t(\mathbf{S}_j \mid \mathbf{D}_i),$$

must increase monotonically as well. This is just equivalent to the average mutual information

$$I_t(\mathbf{S}, \mathbf{D}) = \sum_{i=1}^{n} \sum_{j=1}^{n} \text{pr}_t(S_j \,\&\, D_i) \log_2 \frac{\text{Pr}_t(S_j \mid D_i)}{\text{Pr}_t(S_j)}$$

Q.E.D.

References

Alexander, R. D. (1990). "Epigenetic Rules and Darwinian Algorithms: The Adaptive Study of Learning and Development." *Ethology and Sociology* 11:241–303.
Aunger, R. (2000). *Darwinizing Culture: The Status of Memetics as a Science.* New York: Oxford University Press.
Axelrod, R. (1984). *The Evolution of Cooperation.* New York: Basic Books.
Barrett, M., and Sober, E. (1992). "Is Entropy Relevant to the Asymmetry Between Retrodiction and Prediction?" *British Journal for the Philosophy of Science* 43:141–60.
Bateson, G. (1972). *Steps to an Ecology of Mind.* London: Jason Aronson.
Blackmore, S. (1999). *The Meme Machine.* New York: Oxford University Press.
Bonner, J. T. (1980). *The Evolution of Culture in Animals.* Princeton, NJ: Princeton University Press.
Börgers, T., and Sarin, R. (1994). "Learning through Reinforcement and Replicator Dynamics." Börgers: University College, London; Sarin: Texas A & M University.
Boyd, R., and Richerson, P. J. (1985). *Culture and the Evolutionary Process.* Chicago: University of Chicago Press.
Brodie, R. (1996). *Virus of the Mind: The New Science of the Meme.* Seattle: Integral Press.
Brooks, D. R., and Wiley, E. O. (1988). *Evolution as Entropy: Toward a Unified Theory of Biology.* 2d ed. Chicago: University of Chicago Press.
Campbell, D. T. ([1974] 1987). "Evolutionary Epistemology." In *Evolutionary Epistemology, Rationality, and the Sociology of Knowledge,* G. Radnitzky and W. W. I. Bartley, eds., 47–89. La Salle, Ill.: Open Court.
Cavalli-Svorza, L. L., and Feldman, M. W. (1981). *Cultural Transmission and Evolution: A Quantitative Approach.* Princeton University Press.
Chalmers, D. J. (1996). *The Conscious Mind: In Search of a Fundamental Theory.* New York: Oxford University Press.
Cheney, D. L., and Seyfarth, R. M. (1990). *How Monkeys See the World.* Chicago: University of Chicago Press.
Dahlbom, B. (1993). *Dennett and His Critics.* Oxford: Blackwell.
Darwall, S., Gibbard, A., and Railton, P. (1992). "Toward Fin de Siecle Ethics: Some Trends." *The Philosophical Review* 101:115–89.

References

Darwin, C. ([1859] 1964). *On the Origin of Species by Means of Natural Selection*, E. Mayr, ed. Cambridge, Mass.: Harvard University Press.
Dawkins, R. (1976). *The Selfish Gene*. New York: Oxford University Press.
Dawkins, R. (1982). *The Extended Phenotype*. Oxford: W. H. Freeman.
Dawkins, R. (1986). *The Blind Watchmaker*. Longman Scientific & Technical.
Dawkins, R. ([1976] 1989). *The Selfish Gene, New Edition*. Oxford: Oxford University Press.
Dawkins, R. (1993). "Viruses of the Mind." Journal: Free Enquiry, Summer 1993, 34–41.
Deacon, T. W. (1997). *The Symbolic Species: The Co-Evolution of Language and the Brain*. New York: W. W. Norton.
Dennett, D. C. (1987). *The Intentional Stance*. Cambridge, Mass.: MIT Press.
Dennett, D. C. (1990). "Memes and the Exploitation of the Imagination." *Journal of Aesthetics and Art Criticism* 48:127–35.
Dennett, D. C. (1991). *Consciousness Explained*. Boston: Little, Brown.
Dennett, D. C. (1996). *Darwin's Dangerous Idea: Evolution and the Meanings of Life*. New York: Simon & Schuster.
Dobzhansky, T. (1941). *Genetics and the Origin of Species*. 2nd ed. New York: Columbia University Press.
Dretske, F. (1981). *Knowledge and the Flow of Information*. Cambridge, Mass.: MIT Press.
Dretske, F. (1986). "Misinformation." In *Belief: Form, Content, and Function*, R. Bogdan, ed., pp. 17–36. Oxford: Clarendon Press.
Dretske, F. (1988). *Explaining Behavior: Reasons in a World of Causes*. Cambridge, Mass.: MIT Press.
Dretske, F. (1995). *Naturalizing the Mind*. Cambridge, Mass.: MIT Press.
Fisher, R. (1930). *The Genetical Theory of Natural Selection*. Oxford: Clarendon Press.
Gettier, E. (1963). "Is Justified True Belief Knowledge?" *Analysis* 23:121–3.
Ghiselin, M. (1997). *Metaphysics and the Origin of Species*. Albany: State University of New York Press.
Godfrey-Smith, P. (1991). "Signal, Decision, Action." *The Journal of Philosophy* 88:709–22.
Godfrey-Smith, P. (1994). "A Modern History Theory of Functions." *Nous* 28:344–62.
Godfrey-Smith, P. (1996). *Complexity and the Function of Mind in Nature*. Cambridge University Press.
Gould, S. J., and Lewontin, R. C. (1978). "The Spandrels of San Marco and the Panglossian Paradigm: A Critique of the Adaptationist Program." *Proceedings of the Royal Society of London* 205:581–98.
Hahlweg, K., and Hooker, C. A. (1989). "Evolutionary Epistemology and the Philosophy of Science." In *Issues in Evolutionary Epistemology*, K. Halweg and C. A. Hooker eds., 21–150. Albany, NY: SUNY.
Hamilton, W. D. (1964). "The Genetical Evolution of Social Behavior I & II." *Journal of Theoretical Biology* 7:1–52.
Harms, W. (1996a). "Population Epistemology: Information Flow in Evolutionary Processes." Ph.D. dissertation. University of California, Irvine.
Harms, W. (1996b). "Cultural Evolution and the Variable Phenotype." *Biology and Philosophy* 11:357–75.

References

Harms, W. (1998). "The Use of Information Theory in Epistemology." *Philosophy of Science* 65:472–501.

Hartley, R. (1928). "Transmission of Information." *Bell System Technical Journal* 7: 535–68.

Hazelbauer, G. L., Berg, H. C., and Matsumura, P. (1993). "Bacterial Motility and Signal Transduction." *Cell* 73:15–22.

Heylighen, F. (1998). "The Memetics Community Is Coming of Age." *Journal of Memetics* 2(2): [editorial].

Hofbauer, J., and Sigmund, K. (1988). *The Theory of Evolution and Dynamical Systems.* Cambridge University Press.

Holland, J. H. ([1975] 1992). *Adaptation in Natural and Artificial Systems.* Cambridge, Mass.: MIT Press.

Horgan, T., and Timmons, M. (1992). "Troubles on Moral Twin Earth: Moral Queerness Revisited." *Synthese* 92:221–60.

Hull, D. (1988a). *Science as a Process.* University of Chicago Press.

Hull, D. (1988b). "A Period of Development: A Response." *Biology and Philosophy* 3:241–63.

Hull, D. (1992). "An Evolutionary Account of Science: A Response to Rosenberg's Critical Notice." *Biology and Philosophy* 7:229–36.

Hull, D. (2001). *Science and Selection: Essays on Biological Evolution and the Philosophy of Science.* Cambridge studies in Philosophy and Biology, New York: Cambridge University Press.

Hume, D. ([1748] 1977). *An Enquiry Concerning Human Understanding.* Indianapolis, Ind.: Hackett Press.

Hume, D. ([1739] 1978). *A Treatise of Human Nature.* Vol. 2. L. Selby-Biggs, ed. Oxford.

Jackson, F. (1986). "What Mary Didn't Know." *Philosophical Quarterly* 32:291–95.

Kapur, J. (1994). *Measures of Information and Their Applications.* New York: John Wiley and Sons.

Kim, J. (1988). "What Is 'Naturalized Epistemology'?" In *Naturalizing Epistemology*, Kornblith, ed., pp. 33–56. Cambridge, Mass.: MIT Press.

Kitcher, P. (1985). *Vaulting Ambition: Sociobiology and the Quest for Human Nature.* Cambridge, Mass.: MIT Press.

Kornblith, H. (1994). *Naturalizing Epistemology.* Cambridge, Mass.: Reprinted in MIT Press.

Kuhn, T. S. (1962). *Foundations of the Unity of Science. Vol. 2, No. 2: The Structure of Scientific Revolutions.* 2nd ed. International Encyclopedia of Unified Science, vol. 2, no. 2. Chicago: University of Chicago Press.

Lamb, J. M., and Wells, H. (1995). "Honey Bee (Apis Mellifera) Use of Flower Form in Making Foraging Choices." *Journal of the Kansas Entomological Society* 68:388–98.

Lewis, D. (1969). *Convention.* Cambridge, Mass.: Harvard University Press.

Lewontin, R. J. (1970). "The Units of Selection." *Annual Review of Ecology and Systematics* 1:1–18.

Lynch, A. (1996). *Thought Contagion: How Belief Spreads through Society.* New York: Basic Books.

Lynch, A., and Baker, A. J. (1993). "A Population Memetics Approach to Cultural Evolution in Chaffinch Song: Meme Diversity Within Populations." *The American Naturalist* 141:597–620.

References

Lynch, A., and Baker, A. J. (1994). "A Population Memetics Approach to Cultural Evolution in Chaffinch Song: Differentiation Among Populations." *Evolution* 48:351–59.

Lynch, A., Plunkett, G. M., Baker, A. J., and Jenkins, P. F. (1989). "A Model of Cultural Evolution of Chaffinch Song Derived from the Meme Concept." *The American Naturalist* 133:634–53.

Manson, M. D. (1990). "Introduction to Bacterial Motility and Chemotaxix." *Journal of Chemical Ecology* 16:107–113.

Maynard Smith, J., and Price, G. R. (1973). "The Logic of Animal Conflict." *Nature* 146:15–18.

Millikan, R. G. (1984). *Language, Thought, and Other Biological Categories: New Foundations for Realism*. Cambridge, Mass.: MIT Press.

Millikan, R. G. (1989). "Biosemantics." *Journal of Philosophy* 86:288–302.

Millikan, R. G. (1990). "Compare and Contrast Dretske, Fodor, and Millikan on Teleosemantics." *Philosophical Topics* 18:151–61.

Millikan, R. G. (1993). *White Queen Psychology and Other Essays for Alice*. Cambridge, Mass.: MIT Press.

Millikan, R. G. (1996). "Pushmi-Pullyu Representations." In *Mind and Morals: Essays on Cognitive Science and Ethics,* L. May, M. Friedman, and A. Clark, eds. 145–61. Cambridge, Mass.: MIT Press.

Mitchell, M. (1996). *An Introduction to Genetic Algorithms*. Cambridge, Mass.: MIT Press.

Montague, P. R., Dayan, P., Person, C., and Sejnowski, T. J. (1995). "Bee Foraging in Uncertain Environments Using Predictive Hebbian Learning." *Nature* 377:725–28.

Montague, P. R., Dayan, P., and Sejnowski, T. J. (1996). "A Framework for Mensencephalic Dopamine Systems Based on Predictive Hebbian Learning." *Journal of Neuroscience* 16:1936–47.

Moore, G. E. (1903). *Principia Ethica*. Cambridge: Cambridge University Press.

Moore, G. E. (1962). *Philosophical Papers*. New York: Collier.

Nagel, T. (1974). "What Is It Like to Be a Bat?" *Philosophical Review*. LXXXIII, 4: 435–50.

Nyquist, H. (1924). "Certain Factors Affecting Telegraph Speed." *Bell System Technical Journal*, p. 324, April 1924.

Petrikin, J., and Wells, H. (1995). "Honey Bee (Apis Mellifera) Use of Flower Pigment Patterns in Making Foraging Choices." *Journal of the Kansas Entomological Society* 68:377–87.

Pierce, J. R. ([1961] 1980). *An Introduction to Information Theory: Symbols, Signals, and Noise*. New York: Dover.

Quine, W. (1969). "Epistemology Naturalized." In: *Ontological Relativity and Other Essays,* 69–90. New York: Columbia University Press.

Quine, W. (1969). "Natural Kinds." In *Ontological Relativity and Other Essays*. New York: Columbia University Press. 114–38.

Raff, R. A. (1996). *The Shape of Life: Genes, Development, and the Evolution of Animal Form*. Chicago: University of Chicago Press.

Real, L. A. (1991). "Animal Choice Behavior and the Evolution of Cognitive Architecture." *Science* 253:980–6.

Real, L. A. (1992). "Information Processing and the Evolutionary Ecology of Cognitive Architecture." *The American Naturalist* 140:S108–S45.

References

Rorty, R. (1979). *Philosophy and the Mirror of Nature*. Princeton: Princeton University Press.

Rosenberg, A. (1992). "Selection and Science: Critical Notice of David Hull's *Science as a Process*." *Biology and Philosophy* 7:217–28.

Schuster, P., and Sigmund, K. (1983). "Replicator Dynamics." *Journal of Theoretical Biology* 100:533–8.

Shannon, C. E. (1948). "A Mathematical Theory of Communication." *The Bell System Technical Journal* 27:379–423, 623–56.

Shannon, C. E. (1993). *Claude Elwood Shannon: Collected Papers*, N. Sloane and A. Wyner, eds. New York: IEEE Press.

Shannon, C. E., and Weaver, W. (1949). *The Mathematical Theory of Communication*. University of Illinois Press.

Siegel, S. (1961). "Decision Making and Learning Under Varying Conditions of Reinforcement." *Annals of the New York Academy of Sciences* 89:752–66.

Skyrms, B. (1994). *The Evolution of an Anomaly*. Technical Report MBS 94–22. Institute for Mathematical Behavioral Sciences, University of California, Irvine.

Skyrms, B. (1996). *Evolution of the Social Contract*. Cambridge University Press.

Sober, E. (1984). *The Nature of Selection: Evolutionary Theory in Philosophical Focus*. Cambridge, Mass.: MIT Press.

Sober, E. (1991). "Temporally Asymmetric Inference in a Markov Process." *Philosophy of Science* 58:398–410.

Sober, E., and Barrett, M. (1992). "Conjunctive Forks and Temporally Asymmetric Inference." *Australasian Journal of Philosophy* 70:1–23.

Sober, E., and Wilson, D. S. (1998). *Unto Others: The Evolution and Psychology of Unselfish Behavior*. Cambridge, Mass.: Harvard University Press.

Sperber, D. (1996). *Explaining Culture: A Naturalistic Approach*. Cambridge, Mass.: Blackwell.

Tomasello, M. (1999). *The Cultural Origins of Human Cognition*. Cambridge, Mass.: Havard University Press.

Trivers, R. L. (1971). "The Evolution of Reciprocal Altruism." *The Quarterly Review of Biology* 46:35–57.

Waddington, C. H. (1975). "Evolution and Epistemology." In *The Evolution of an Evolutionist*, 35–6. Edinburgh University Press.

Weber, B. H., Depew, D. J., and Smith, J. D. (1988). *Entropy, Information, and Evolution: New Perspectives on Physical and Biological Evolution*. Cambridge, Mass.: MIT Press.

Wheeler, J. A. (1994). "It from Bit." In *At Home in the Universe*, 295–312. Woodbury, N.Y.: American Institute of Physics Press.

Wicken, J. S. (1987). *Evolution, Thermodynamics, and Information: Extending the Darwinian Program*. New York: Oxford University Press.

Wicken, J. S. (1988). "Thermodynamics, Evolution, and Emergence: Ingredients for a New Synthesis." In *Entropy, Information, and Evolution*, B. Weber, D. Depew, and J. Smith eds. 139–69. Cambridge, Mass.: MIT Press.

Wiener, N. ([1948] 1961). *Cybernetics, or Control and Communication in the Animal and the Machine*. 2nd ed. MIT Press.

Williams, G. C. (1966). *Adaptation and Natural Selection*. Princeton: Princeton University Press.

References

Wilson, E. O. (1975). *Sociobiology: The New Synthesis*. Cambridge, Mass.: Harvard University Press.
Wilson, E. O. (1979). *On Human Nature*. New York: Bantam.
Wittgenstein, L. (1953). *Philosophical Investigations*. New York: Macmillan Company.
Wright, S. (1932). "The Roles of Mutation, Inbreeding, Crossbreeding, and Selection in Evolution." *Proceedings of the Sixth International Congress of Genetics* 1:356–66.
Wright, S. (1986). *Evolution: Selected Papers*. Chicago: University of Chicago Press.
Zurek, W. H., ed. (1990). *Complexity, Entropy, and the Physics of Information: Proceedings of the SFI Workshop*. Santa Fe Institute Studies in the Sciences of Complexity, vol. VIII. Redwood City, Calif.: Addison-Wesley.

Index

adaptationism, 56
adaptive landscape, 89, 140, 144, 166
adaptive plasticity, see phenotypic variability
adaptive target, 141
aquaintance, knowledge by, 220

beaver tail slaps, 119, 121, 137, 204
bee dance, 120
belief/desire distinction, 201, 225
bits, 114, 115
Bateson, Gregory, 115
Blackmore, Susan, 66
Brodie, Richard, 65
bumblebee model, 164–180

Campbell, Donald T., 4, 33, 47, 72, 85, 94, 151, 165, 169
cellular ontology, 54–57
Chalmers, David, 115
Churchland, Pat, 223
cheater detection, 209
co-adaptation, 197, 198, 200
common knowledge, 196
common sense, 4, 159–161, 163
comprehension, 213
computer modeling, see modeling, computer
consciousness, 42–43, 44
content, 194
  primitive, see primitive content
conventions, 194, 195, 200, 203, 211, 231, 232–233
  enforcement of, 206, 207, 208, 209, 214
  formalization of, 203–204, 208
  normativity of, 204, 233
  result of selection, 196
cooperation, evolution of, 100–102

copying, 19, 23, 49, 55, 64
  error, 21, 23
correspondence, 64, 200, 203, 204, 205, 206, 207, 208, 232, 237
cultural evolution, 8, 50, 151, 234
  adaptive, 64
  defined, 64
cultural vs. genetic determinism, 44, 150, 219, 231, 240
  requires replicators, 28, 31, 40, 41, 48, 65, 82
curiosity, 171

Darwall, *et al.*, 221
Darwin, Charles, 17
Dawkins, Richard, 8, 16–28, 49, 64
demes, 106
  in science, 35–38
Dennett, Daniel, 8, 64, 72
development, 54, 57
discreteness
  of genetic code, 21, 23, 24, 49
  of memetic code, 25, 26, 131
disjunctive referents, 117, 199, 204
dispositions, 39, 61, 62–63, 83, 148
  higher-order, 62
Dobzhansky, T., 17
Dretske, Fred, 109, 116–118, 135, 201
drift, see sampling error

E. Coli, 153, 206
empiricism, 135
  ahistorical bias, 5, 6, 201, 229, 230
epistemology, 3–4, 189
  evolutionary, 4–6, 40, 59, 85, 185, 186
  naturalistic, 4–6, 189
  problem of, 133–138, 162, 182, 183, 184

265

Index

error detectors, see success indicators
evolution
  cultural, see cultural evolution
  defined, 29, 47, 71
  standard introduction, 81
  transformative, 73
exaptation, of success indicators, 173, 181
extension, 1, 193, 204, 212, 226
extinction, 107

fertilization, 57
Fisher's Fundamental Theorem, 36, 142
fission, 55
fitness
  conceptual inclusive, 33–35
  defined, 88, 91, 92
  frequency-dependent, 99
  inclusive, 34
  mean, 89
fitness feedback loop, 85, 164, 184
foundationalism, 5
functions, biological, 117
function-stabilizing mechanisms, 202

gap
  is/ought, 6, 190, 219, 224, 225, 227
  Kantian, 134, 140, 147, 162, 183, 184
game theory, evolutionary, 9, 73, 75, 86, 195, 196
gene, "sliding definition" of, 21, 24, 25, 41, 45, 46, 49, 53
genetic algorithm, 85
Gettier, Edmund, 210
Godfrey-Smith, Peter, 109, 118, 121
gradualism, 17, 20, 71

Hamilton, William D., 34, 207
Heylighen, Francis, 64
historical relations, 7, 192, 197, 200, 219, 230, 231
Holland, John, 85
homeostatic mechanisms, 60–61
Horgan, Terry, 236
Hull, David, 8, 28–41, 106
Hume, David, 5, 6, 9, 10–12, 160, 211, 219, 228–231
hunger, 205

identity conditions
  of gene, 18, 41
  of memes, 24, 26, 45, 46, 49
  of replicator, 20, 39
imitation, 24, 63, 83
information, 25, 45, 60, 129, 154, 158
  additivity requirement, 113
  as "difference that makes a difference," 115
  channel conditions, 117
  current mutual, 112
  entropy, 110, 185
  equivocation, 112, 116
  exploitability of, 146, 147, 158, 174, 175, 180, 233
  individual level, 146
  instructional, 131
  mutual, 112, 114, 115, 118, 122, 131, 143, 147, 173, 185
  population level, 145
  rate of transmission, 111
  self-, 110
  semantic, 116, 117, 130, 132
  spaces, 115, 131
  structural, 131, 162
  symmetry of, 112
  utility of, 123, 124, 126–129, 148
innate traits, 152, 160, 161, 168, 215
intension, 194, 211
intentionality, 193
intentional icon, 199
intentional properties, 43, 45, 46, 194
intentional stance, the, 45–46, 50
interactor, 29, 30–31, 33, 40
internal states, 164
intuitions, 210–211, 212–213, 220, 221, 234
  authority of, 7, 190, 209, 235
  conflicts, 237, 238, 239, 240
  meaning of, 192, 205, 214, 235

Jackson, Frank, 223
justification, 4, 120, 204, 211

Kim, Jaegwon, 6, 235
Kuhn, Thomas, 2

Lamarkian inheritance, 26, 70
learning, 150, 165
  conservativeness in, 172, 177
  in populations, 138, 180
  trial-and-error, 84, 151, 160, 185
Leibniz, Gottfried Wilhelm, 10
Lewis, David, 195
Lewontin, Richard, 47

Index

lineage, 30, 39
  defined, 29
logarithm, 114
logic, 213

Mary problem, 223
meaning, 4, 7, 195, 196, 197, 200
meme, 23–28, 32
meme's-eye view, 44, 64, 66, 67, 75
memetics, 59, 64–67, 75, 231
memory, 160
Millikan, Ruth, 9, 118, 191, 194, 197–198, 200, 212, 214
misrepresentation, 117, 120
modeling, 102, 185
  arbitrary parameter values, 173
  in evolutionary theory, 17
  computer, 9, 85–86, 145, 165
  vectors, 87
Moore, G.E., 7, 9, 161, 219
motivation, direct, 191, 201
mutation, 95–99, 104, 108, 171
  matrix, 99
  selection-driven, 97, 168
  uniform random, 95, 145

naturalistic fallacy, 220
normalization, 89, 91, 104, 171
normative theory, 6, 190, 225
norms
  ethical, 4, 5, 191, 209, 210, 219
  epistemological, 4, 5, 189, 190
  incommensurability, 190, 227, 238
  participation in, 225, 226, 237

ontological neutrality, 114, 119, 132, 136, 141, 147, 181, 183, 185, 203
open-question argument, 9, 220, 227

partition, 143
  by behavioral options, 119, 204
phase portrait, 88
phenotypic variability, 62, 153, 162, 172, 185
physics, analogy with, 74
population genetics, 9, 73, 86
populations, 86, 94, 174
  defined, 87
  effectively infinite, 36, 105–107
  evolution of, 73
  fixed-size, 104
practices, transmission of, 32, 39

preferences, 169
primitive content, 201, 202, 214, 216, 218, 225, 233
  and antinaturalism, 222, 230, 234
  directly motivating, 205
  in regulatory hierarchies, 211, 215
  limits of, 237
  formalized, 203–204
  nonpropositional, 201
  prisoner's dilemma, 101, 142, 207
  propositions, 7, 191, 201, 205, 214, 215, 218, 234, 235

Quine, W.V.O., 5, 189, 201

random walks, 154, 164, 168, 174
Real, Leslie A., 164
reciprocal altruism, 207
recombination, 57
regulatory hierarchy, 206, 207, 209, 211, 214
relativism, 7, 28, 36, 38, 228
  epistemological, 1, 135
reliability, 120, 122, 130, 215
  of senses, 4, 151, 183, 233
replication, 54
  self-, 8, 19, 47, 55, 56
  see also copying
replicator
  active/passive, 19
  definition of, 18–23, 29
  germ-line, 19
  generality of, 22
  longevity of, 20
  summary of properties, 23
  test of mutilation, 21–22
  see also gene
replicator dynamics, 15, 87–89, 171, 185
representations
  consumption of, 119
Rorty, Richard, 3
Rosenberg, Alex, 39
rules, see conventions

sampling error, 36, 68, 136, 140, 145
selection, 4, 31, 88, 91, 93, 136
  defined, 29, 89
  levels of, 29
  local vs. genetic, 155
  multilevel, 149, 151, 165, 169, 185
  units of, 17, 20, 29, 53

Index

selection vs. variation
  causal criterion, 93
  mathematical criterion, 93–94
Selfish Gene, The, 16–17
selfishness, 44, 47, 49, 54, 65, 68–72
  self-beneficial, 68, 69
  economic, 68, 69–70
  accumulation of traits, 68
Shannon, Claude, 109, 113
signaling system, 195
sociobiology, 66, 161
stability, 18, 81, 136, 139
stimulus meaning, 199
success indicators, 84, 151, 158, 159, 160, 161, 163
supervenience, 50, 235–237
simplex, 87
skepticism, 135
Skyrms, Brian, 9, 107, 195
Sober, Elliott, 106
  and D.S.Wilson, 53, 69–70
structure, transmission of, 30
symmetry breaking, 196

tautology problem, 68
teleosemantics, 109, 194
  alternative foundations, 215–216, 233
  formalization of, 203–204, 236
  see also primitive content
Tit-for-Tat, 101
tracking efficiency, 109, 112, 120–122, 185

translation, 226, 227
transmission, cultural, 8, 55, 62, 63
  see also evolution, cultural
triangulation, 134
Trivers, Robert, 207
truth, 4, 163, 211
  see also correspondence
  and information transfer, 91, 127, 140, 144–145, 146, 183
type-token distinction, 18, 30, 41
type scheme, imposition of, 139, 183

uncertainty, 110, 113

variation, 92, 93, 97, 145, 180
  biased, 169
  rate of, 23, 70, 97, 177
vehicles, 33
vervet monkeys, 198, 200, 226, 238
vicarious selectors, 84, 151, 152, 184
  see also success indicators

warning cries, 7, 200
Waddington, C.H., 160
Wiener, Norbert, 113
Williams, George C., 21
Wilson, E.O., 161
Wittgenstein, Ludwig, 216
word, as unit of meaning, 192, 201, 212, 214, 215, 217, 226
Wright, Sewall, 36, 89, 106